PEOPLE
AND
ENVIRONMENT

Edited by
Stephen Morse
Michael Stocking
School of Development Studies
University of East Anglia

Routledge
Taylor & Francis Group

LONDON AND NEW YORK

First published 1995 by UCL Press
Published 2013 by Routledge
2 Park Square, Milton Park, Abingdon, Oxon OX14 4RN
711 Third Avenue, New York, NY 10017, USA

*Routledge is an imprint of the Taylor & Francis Group,
an informa business*

ISBNs:
1-85728-282-5 HB
1-85728-283-3 PB
ISBN 978-1-857-28283-2 (pbk)

British Library Cataloguing in Publication Data
A catalogue record for this book is available from the British Library.

Typeset in Plantin and Optima.

CONTENTS

EDITORS' FOREWORD

"Humanity must live within the capacity of the Earth. There is no other rational option in the longer term", says *Caring for the Earth* (IUCN/UNEP/ WWF 1991). Yet the 1980s and 1990s have witnessed abundant evidence that the planet Earth is under stress and that people are not recognizing the "rational" option; environments are degrading; natural resources are being depleted faster than they are being renewed; and populations, especially in the developing world, are growing faster than exploitative technologies can meet needs, let alone wants. The carrying capacity of at least parts of the Earth's surface is being grossly exceeded. At the same time, social, political and economic strife exacerbate the relationship between humans and their environment, to the extent that the Earth is put under further pressure. In such a milieu – where poor, desperate and hungry people exist alongside the affluent and powerful – development has come to be seen as the means of transferring capital resources and stocks to the needy.

Reaching the homes of most persons in aid donor countries, television and media attention have highlighted the human suffering in places such as Ethiopia, Rwanda and eastern Europe. Less immediately obvious are the pressures exerted by the rapidly expanding economies of the Pacific Rim and the newly industrializing countries (NICS), and how these have changed lifestyles for millions. Every human being has foremost the desire to survive, after which comes the need to improve livelihoods, and eventually the wish to satisfy wants. However, because of the way we live, especially the way that the 1000 million in the affluent West live, the life-support systems for the nearly 6000 million of the world are at risk through overstressing the Earth. This is the challenge of the relationship between "people" and their "environment" – how can the circle be squared? How can a finite Earth, with limited renewing c#apacity in its systems of natural resources, provide for a growing population with increasing aspirations. Part of the answer lies in society, how it is structured and how issues of equity are addressed; part rests with technology and access to resources; and part must tackle questions of economic growth, control of consumption and demands by populations. For environmentalists, the only lasting answer is the development of a new ethic for sustainable living, where development is in the service of both nature and humans. This book addresses some of the fundamental facets of these complicated relationships between people, their environment and development.

International conferences such as the Earth Summit at Rio and the Cairo Population Conference along with international accords for global trade (GATT) and trade in endangered species (CITES) have helped to attach these complex interrelationships to the agendas of most countries and most people. The "global village" and its "global commons" are no longer romantic pleasantries but expressions of the hard issues faced by humanity as resources per capita decline. Famines in Africa are as real to most people as unemployment in Scunthorpe: both are evidence of breakdown in the capacity of human societies to deal with complex socio-economic and environmental problems.

This book tackles issues surrounding the theme of development from several perspectives: global environmental change, political economy, gender analysis, technological and conservationist. It has been edited and written by faculty and staff at the University of East Anglia, most of whom work in the internationally acclaimed School of Development Studies (DEV) and its Overseas Development Group (ODG). ODG was formed in 1967 from a group of concerned sociologists and economists at the university, with the specific remit of enabling academic teachers and researchers to practice what they preach and of benefiting their students with practical experience of development matters. Subsequently, the teaching School, DEV, was created in 1973 to build a coherent set of teaching programmes around the theme of development and developing countries. DEV is still unique in the United Kingdom: it has a full range of teaching from diploma and undergraduate programmes, through nine masters programmes and onto a lively research grouping for MPhil and PhD degrees; it is the only department of its kind with a significant attachment to scientific, technological, environmental disciplines in its teaching and research; all other establishments specialize in the economic and social sciences, which, although obviously crucial, lead only to a limited set of analyses of the problems of development and possible solutions. Through the ODG, all faculty are also practitioners, leading to a degree of pragmatism unparalleled in competitor institutions. These commitments to multidisciplinarity, interdisciplinarity and actual practice help to provide a range of insights and dimensions that would be impossible in a standard academic institution.

The School has long sought to help promote development issues beyond the immediate student population, and it is for this reason that a series of public lectures is held in the autumn of each year, to which students, university staff and the general public are invited. The chapters in this volume derive originally from the lecture series of 1990 around the theme of "People and environment: development for the 1990s". Subsequent debate and critical analysis have led to a refinement over three years of the messages of those lectures, to the extent that some of the chapters here are barely recognizable from their oral presentations. In addition, one additional contribution has been sought to set the population–environment debate in a theoretical political economic (or ecological) context for which the School is well known.

This book aims to promote accessibility to current debates in development, spanning the natural and social sciences. It is intended as a primer for undergraduate and postgraduate students who are attempting to get a feel for these issues for the first time. The topics covered are intended to provide the reader with a broad sweep of issues and up-to-date access to the literature on subjects such as sustainable agriculture, gender analysis, population growth and biodiversity, as they relate to developing countries. However, a book of this type inevitably falls short of covering all topical issues in development. This is in the nature of the subject: a spotlight on development and development processes as the "stage", in which the "actors" (topics such as biotechnology) play out their roles, speak their lines and interact with the villains. Not all the stars are on this stage, and the reader will no doubt recognize some who are missing from the play. We cannot guarantee a happy ending – but the plot should be interesting

The opening chapter deals with human perceptions of the environment, and argues cogently for a wider political analysis of the use of natural resources and the environment. This develops Piers Blaikie's well known "chain of explanation" and his discourse on the *Political economy of soil erosion in developing countries* (1985). His contribution in this book provides a useful deconstruction of society as it engages the population–environment interchange.

The second chapter disentangles the complex, and somewhat confusing, concept of sustainability. The concept is sometimes criticized for being all things to all people. Hence, it is often difficult for students to access, not least because of the now vast literature that has radiated from it. David Gibbon, Alex Lake and Michael Stocking provide an overview of the various ways that sustainability is perceived and how the concept provides a significant challenge to modern agriculture and entrenched views of production maximization. We thought that "sustainability" would be a passing fad, a jargon word for the 1980s, but it has proved more durable and lives on in several important and lively debates such as "indicators of sustainability" and changing lifestyles to accommodate our future and the survival of the "global commons". These debates will run and run!

Environmental change and its relevance to development is the theme of two chapters in this book. The first, by Mick Kelly and Sarah Granich, takes a macro perspective of potential impacts that may arise from human mediated global warming. The "greenhouse effect" resulting from the release of gases such as carbon dioxide and methane has been the subject of much concern in recent years, and the authors use estimations from the latest research in this field to explore the range of environmental change possibilities that we may one day have to face. The second contribution comes from Cecile Jackson, who takes a very different viewpoint on environmental change. Instead of considering potential impacts of change as socially neutral (i.e. the same for all), the opposite is stressed: environmental change hits some harder than others, and the worst affected are usually the weakest and most disadvantaged in society. In this chapter, gender differentiation is highlighted, a perspective from which many development issues are now being revisited.

History is replete with examples of "technical fixes" being promoted by the West as answers to underdevelopment. The Green Revolution was one such approach, and the echoes reverberate down to the present with the heralding of a new technical champion – biotechnology. This field is replete with rhetoric, and the fifth chapter by Stephen Morse critically examines potential impacts (good and bad) that are often trumpeted for biotechnology by scientists and their detractors.

The desire for development is often seen as inimical, and indeed diametrically opposed, to the need for nature conservation. Trade-offs between the two have formed the basis for a debate that in one form or another has exercised humans ever since Plato first wrote of the impact of humans on nature. The arguments have, of course, become much more defined (and immediate) in recent decades, as the pace of human encroachment into "natural" habitats has increased. The chapter by Mike Stocking, Scott Perkin and Kate Brown summarizes this debate with particular emphasis on two projects in Africa that have tried to provide a practical resolution to what most perceive as the conflict between biodiversity conservation and economic development. Maybe it does not have to be a conflict, the authors argue.

Human population growth has been so linked with development in the eyes of many that the popular perception is now one of an inevitable and reciprocal relationship between the two – adequate development being reliant upon reduced population growth. Ian Thomas, in the final chapter, critically examines this assumed linkage, and draws some lessons for the future.

Given the broad coverage of topics in this book, and the assumption that readers are more likely to "dip" into chapters rather than read them in order, the editors have decided to provide an introduction to each chapter rather than write a separate introductory chapter. The introductions will explain the reasons that lay behind the inclusion of the topic in the book, and provide a context to help the reader see the linkages with other topics that are covered. The unifying theme of these introductions is *Agenda 21*, the most tangible of the outputs from the "Earth Summit" at Rio de Janeiro in 1992, where for the first time the majority of the world's leaders supported the general concern for environment-development linkages – but, of course, many of their individual agendas were different. Now, post-Rio, we can see that the Summit was a milestone in raising the profile for taking seriously the impact of people on their environments. This book is, hopefully, a further contribution to the clarification of the dominant issues, and the construction of a research and teaching agenda to debate the challenges.

Stephen Morse   Michael Stocking

Norwich, November 1994

CHAPTER ONE

Understanding environmental issues

Piers Blaikie

Editors' introduction

Too often, simplistic answers to perceived environmental issues are provided from single and narrow perspectives. Our understanding of these issues rests partly on detailed and specialized research carried out within a single discipline or subdiscipline in universities and research establishments. The problem arises when we allow ourselves to believe that this is the *only* source of understanding. Environmental issues are by definition also social ones, and therefore our understanding must rest upon broader interdisciplinary perspectives that transcend institutional and professional barriers.

However, there will always be tensions between narrow but powerful conceptualizations of environment and society (typically these have come from the natural sciences and econometrics) on the one hand, and broader interdisciplinary views about the interactions between people themselves, and between them and the places in which they live. The structure of rewards and penalties of professional life in universities, multilateral institutions and policy-making bodies have unfortunately tended to underlie these tensions to produce antagonistic outcomes – which reproduces the single and narrow perspectives with which we started out.

This book develops an approach that embraces multiple perspectives and attempts to make sense of their differences and commonalities. These are not only derived from within different academic disciplines of both the natural and the social sciences, but from others outside the policy-making and academic environment altogether. For example, this book has perspectives that examine the often conflicting needs of development and conservation (Ch. 6), from those who view environmental change through a gender lens

(Ch. 4), and from those concerned with the sustainability of development (Ch. 2). Together they form a plural and complex view of environment–society relationships. These are essential (but often unheard) voices giving us local understandings and articulating local interests regarding their different environments.

This chapter highlights these constraints and opportunities to understanding the complex relationships between environment and society. Piers Blaikie first draws our attention to the political and social aspects of constructing a natural-science view of the environment. He then goes on to take a multilevel approach to environment–society relationships called "political ecology", an interesting hybrid label that describes something of our approach in this book. Having established a methodology that links different levels and scales, he illustrates it by taking the case of land degradation, which provides both a substantive contribution to how we understand this environmental and social problem, and an illustration of a methodology with wider applications.

Introduction

Much of this book is about global environmental issues, and is written for the most part by natural scientists. However, how we may define what is an "environmental issue" is not as straightforward as it may seem, and is not a matter for natural scientists alone. This chapter will explain why environmental issues are also social issues, and therefore they tend to lie outside the usual scope of the natural sciences themselves. Therefore, there are several questions of epistemology (or the theory for the method for analysis) and of approach in more general terms, which both social scientists and natural scientists have to negotiate. There are three main reasons why environmental issues are also social.

First, the ways in which changes in the tangible and "real" physical environment impinge upon peoples' lives have to be interpreted by people. They have first to be experienced by them (and by different people in different ways), become objects of thought, and then articulated. This may be done by scientists, politicians and by a great variety of other citizens too – whether they are artisanal fishermen off the coast of Brazil, women collecting firewood in India, or medical staff coping with pollution-induced illnesses in the industrial areas of the Czech Republic. In some cases of experience of environmental change, such as pollution, long-term land degradation and other insidious and unrecognized impacts of environmental change, explanations linking symptom with cause are simply not in place, and it remains unperceived.

Secondly, environmental issues require a vehicle for their political expression to enter discourses at the local, national or global levels. They have to resonate with interests of political communities, otherwise they may remain at the level of experience only (often unpleasant, dangerous and impoverishing), without ever becoming an issue. This means that some environmental changes that impinge on peoples' lives never reach the surface of public life, with an airing of differences of view, public debate and politicization. Those that make the public arena are often those that resonate with other continuing political discourses, which may not originally concern the environment at all.

For example, soil conservation was a major concern of many colonial governments in Anglophone Africa and Asia, and as such was an environmental issue, since it was diagnosed by government scientists and incorporated into agricultural policy and extension. However, the colonial State applied various forms of coercion to indigenous farmers to carry out conserving practices, which were resented. These measures reinforced other resentments about interference in rural Africans' lives, and about the fact that, in settler States, their land had been expropriated, causing instant "overpopulation" and the very environmental problems identified by the colonial State. Thus, an environmental issue (in the form of State-coerced conservation programmes) became a marker and focus for resistance for independence movements (Fosbrooke & Young 1976). Today, even after more than a generation of post-colonial experience, official soil conservation is still linked to colonial oppression and government interference, and has therefore become very much reduced in importance as an environmental issue at the national level throughout much of Anglophone post-colonial Africa).

Thirdly, the ways in which environmental issues are understood and enter discourse will reflect the different actors involved. There will probably be scientists from different disciplines with varying foci on separate aspects of the social–environmental system. They will have different notions about the status and domain of what constitutes evidence and proof. Also, there will be a variety of others who are experiencing these environmental changes in different ways. For example, women in the South often experience these changes in a more immediate and severe manner than men (see Jackson, Ch. 4). In more general terms, the more wealthy can probably mitigate the physical symptoms better than the poor, and be able to bring to bear their interpretations and interests more forcibly in the political arena. There will be direct economic interests that will become involved in both the changes and in the measures promoted to deal with them. They may be lobbies for the coal or nuclear industries and biotechnology firms, which may resist measures to regulate access to genetic resources, or poor settlers who will lose their livelihoods in the event of restrictions on using tropical forests. It was clear that these powerful interests were brought to bear on the national negotiating teams at the recent UNCED World Summit and also in the most recent round of GATT negotiations. There will be a varied cast of actors in any environmental issue as it is played out on a variety of political stages.

This introductory chapter will now examine the implications of these social aspects of environmental issues, and will identify the task of engaging with the variety of views about them. This will involve a framework for a holistic and interdisciplinary understanding, which bridges natural and social sciences, and scientific and other "nonprofessional" interpretations.

Nature, environment and society

At the outset let us distinguish nature from environment. "Nature" is not related to or conceptually centred upon something, such as an organism or humankind in general, but "environment" certainly is. There cannot be an organism without an environment, and likewise there cannot be an environment without an organism. After all, nature as opposed to the environment (Einstein defined it as "everything except me") continues to exist without the conscious attention of "me" or indeed humankind at all. For this reason it is difficult to argue that environmental issues somehow simply exist out there in nature without either human agency or human recognition and definition. It is by a process of recursive interaction with the environment that we perceive and act within and upon it. Thus, the environment has on offer, as it were, what we may call "affordances", or inherent qualities that are offered and provided to organisms, including human beings. In the latter case they become opportunities and constraints in the production of use values, the materials for which are "afforded" by the environment. When these affordances are taken up by humans, natural elements in nature become resources in the environment. This process is summed up in the saying that "natural resources are not, they become". In the same way, the sink and service aspects of the environment can also be thought of as affordances, which are on offer from the environment. Thus, an environmental issue is caused by socio-economic actions that in some way is felt to be detrimental to these affordances, actual or potential. It is perceived by people in particular ways, and it enters (or fails to enter) the public arena through the political process. In other words, someone somewhere is not availing themselves of the affordances to which they are used or feel themselves entitled, and are powerful enough to put this on someone else's agenda.

It follows, therefore, that there are preconditions before these objective changes become so-called "issues". To begin with, certain aspects of the environment have to become objects of thought, and do so mainly because they are involved in the activities of production and consumption by society. For example, as Chapter 3 will explain, it was not until the 1890s that a Swedish scientist forecast that coal burning would release enough carbon dioxide into the atmosphere to change the global climate. Thus, the rate of use of a natural resource with existing technology has exposed a limitation in other affordances in the environment (for example, the expectation of stable sea level and related risks of flooding and storm damage in coastal areas). However, this does not necessarily mean that it amounts to an environmental issue. So, the two preconditions may be summarized as:

- recognition by civil society, either through experience, which is expressed by scientific identification or other means
- the political platform to promote this experience.

Both are necessary conditions, but no one of them is sufficient, and they challenge the conventional assumption that it only takes objective scientific facts to frame innate problems "out there" in the environment, thereby defining and creating the issue by scientific discovery alone.

The environment has always been politicized. From earliest history, humans have been in competition over the definition and use of natural resources – in a sense implying the evolution of those struggles from notions of ecological competition between organisms within a habitat to a political economy of environment within society. A central element in the history of the development of capitalism has been the definition of objects in nature as natural resources, subject to global struggles between imperial powers and between them and indigenous peoples. In this century, the environment has become even more consciously politicized for several reasons.

Increased pressure on the environment

The accelerating development of capitalism with economic growth of the North and the newly industrializing countries of the South, and population growth of the South, have created increasingly acute scarcities that are highly differentiated by world region, class, gender and ethnicity. These scarcities of natural resources, and the overloading of the abilities of the environment to disperse noxious wastes harmlessly, have become more severe, and are recognized as such by a variety of actors. This includes the struggles of rubber-tappers in Amazonia, or of poor peasants in central India against the flooding of their homes for hydroelectric schemes, to salubrious conference venues at which scientists discuss global biodiversity loss.

The growth of "environmentalism"

The environment has become subject to a more sustained and coherent reflection by writers and activists predominantly from the North, which can be encapsulated as "environmentalism" (O'Riordan 1981, Pepper 1984, Eckersley 1992). Philosophical, ethical and political concerns have enriched our understanding of the relationships between environment and society. These concerns have also provided a basis for concerted political action, again mostly in North America and western Europe (in the form of Green parties), but also by activist groups in the South, especially in India and Latin America.

Global perspectives on the environment

Following the other two developments mentioned above, the environment has become since the early 1970s the subject of global discourse, culminating in the first international conference on the theme (the United Nations Conference on the Human Environment in Stockholm in 1972). As has been noted by Ensenberger (1974), it is only when environmental change affects adversely the politically powerful North that the issue becomes "global". Otherwise, for those affected people who do not have the power to pack a powerful knowledge claim in the international arena, the issue remains local, and outside global discourse altogether. However, many such issues have stimulated international attempts to address global environmental problems. These include the:
- (aborted) International Law of the Sea
- International Whaling Commission (IWC)
- London Dumping Convention (now called the Dumping Convention).
- Convention of International Trade in Endangered Species (CITES)
- Vienna and Montreal Protocols dealing with the elimination of substances harmful to the ozone layer
- UNCED at Rio in 1992 with its Conventions on biodiversity and global warming.

These examples along with a host of other regional environmental accords, is witness to the fact that the environment has become part of international politics.

As the study of the environment has become extraordinarily complex, contradictory, multi-stranded and multi-disciplinary, what implications does this have for how writers, academics, scientists and policy-makers understand it? This introductory chapter examines the problems and challenges raised by this question, and provides an outline framework in which different analysts and others involved in any way with the issue at hand can negotiate their differences in understanding their epistomologies (that is to say, the different theories they use in the method of understanding and analysis), and their disciplinary approaches. Even the briefest overview will identify the following disciplines: anthropology, biology, botany, climatology, ecology, economics, geography (physical and human), geomorphology, legal studies, political economy, political science, sociology, soil science and zoology. A complete list would be tedious, but would also have to include a range of hybrid studies too. Each of these disciplines, as practised, has different approaches, epistomologies and fields of interest. It is therefore important that they do not "talk past each other" and fail to engage in constructive dialogue about the technical and social aspects of environmental issues. Some ways in which this might be achieved are discussed in the next sections.

The interface between these academic fields and the arena of policy-making is one increasingly recognized as both vital but also problematic to the solution of environmental problems. One of the major issues is the use to which scientific facts are put, what ones are used, and what fails to make any significant agenda. The increasingly dated rationalist view of science – that policy-makers simply use the facts of objective science – is difficult to sustain, since it has been increasingly recognized that scientific information, as "authoritative knowledge" is frequently used selectively to legitimize particular policies. Thus, the view of science in policy-making as "truth talking to power" still raises unanswered questions. That science will, independently and in an apolitical world, uncover what the environmental problems are and will tell the policy-makers what should be done, invites questions about how and why knowledge is created, promoted and used.

Political ecology: bringing environment and society back together

To address these daunting tasks, this chapter uses a perspective derived from the developing field of "political ecology". As its name implies, it is a hybrid field that includes both the natural and the social sciences, and as such it confronts some central challenges in understanding society and environment. Political ecology brings together the diverse disciplines and approaches listed above, and internalizes them in both oppositional and eclectic ways. Early uses of the term by Wolf (1972), Ensenberger (1974) and Cockburn & Ridgeway (1979) drew attention to the increasing politicization of the environment in ways suggested above. Clearly, politics and ecology were involved, but also three other approaches were implied:

* political economy
* human ecology (discussed further in this volume by Stocking)
* cultural ecology.

Since the 1970s, political ecology has rather taken on the role of the wheel – it is reinvented again and again, underscoring that political ecology continues to be a useful and necessary idea for the understanding of environment in a contemporary, multi-disciplinary context. Blaikie & Brookfield (1987) developed the notion at length, and the late 1980s and 1990s are replete with further references (e.g. Bassett 1988, 1993, Bryant 1992, Colchester 1993). For example, Zimmerer (1993) has identified five different ways human geography has used ecological ideas, of which political ecology is the most recent. Each of these approaches within a broadly defined ecology, as well as those from political economy, themselves have an enormous diversity of views. The task of the development of political ecological understanding of people and environment may be one of:

* identifying productive contradictions between and within its parent disciplines
* ordering these contradictions into a set of mutually interacting discourses
* providing a negotiating space for the contradictory views and approaches that politicized environmental studies requires.

Intellectual growth has often been strongest at the antagonistic margins of established disciplines and their institution-bounded modes of production. These margins are constantly shifting, sometimes incrementally, sometimes decisively (after Kuhn's paradigmatic shifts: Kuhn 1962). *Therefore, political ecology may itself have a useful half-life of several years,* during which it brings together hitherto unopposed ideas, to produce qualitatively new ways of understanding.

8

Negotiations about epistemology

A starting point implied in political ecology is the active engagement of natural and social sciences in understanding the production and reproduction of nature and society. Peet & Watts (1993) have characterized the project as "reflect[ing] a confluence between ecologically rooted social science and principles of political economy" (Peet & Watts 1993: 239). Bryant (1992: 120) identifies the need for a political ecology on the grounds that "environmental and political forces will mediate Third World development in unprecedented ways", whereas Blaikie & Brookfield (1987: 17) define it as "combining the concerns of ecology and a broadly defined political economy". Although these locations of the subject are unnecessarily narrow, they do identify the necessary bringing together of natural and social science. In so doing they combine differing epistomologies, which have to be negotiated in the holistic understanding of contemporary environmental issues.

On the one hand it is necessary to understand the natural processes involved in the production and reproduction of natural resources (what Pepper (1984: 60) calls the "real" or tangible physical environment), but be aware that all this has to be perceived by people themselves through cultural filters: the point made at the beginning of this chapter. These are not arrived at independently from the environment, as conventional cultural ecology implies, but by recursive experiences with the environment and with each other. As White (1967) put it: "What we do about ecology depends upon our ideas of the man–nature relationship". Pepper goes on (op. cit.: 7) to quote Pryce (1977) "from time immemorial people have formed opinions, developed attitudes and based their actions on images that may have borne little or no resemblance to reality. Biassed or indeed, completely erroneous ideas concerning the environment are potentially just as influential as those conforming to the real world." This introduces the possibility that these ideas may become partially or completely separated from what is actually happening in the real and tangible environment. Clearly, "environment" sometimes becomes an object of thought and a focus for struggles over meaning and ideology, and is separated from the detail of physical changes in the environment. The case of charismatic species, such as the whale, lion or rhinoceros, illustrates how far images can be substituted for a complex and contradictory reality. The exhortation "save the whales" privileges them above other (even more) endangered species, and usually does not differentiate between different species of whale.

The position taken here is that, although the environment must be considered to be socially constructed, there are aspects of the environment that are more open to alternate interpretations and therefore can be said to be more socially constructed than others. This means that some are closer to "facts" and a single and unopposed interpretation from a variety of actors than others. In this volume there are examples across the continuum. For example, both Morse's chapter on biotechnology and that of Stocking et al. on conservation touch on the issue of biodiversity. However, the implicatións for biodiversity loss can be viewed in many different ways. This can be illustrated by taking the viewpoint of local resource users and international firms.

- Local resource users in rural area of the South will suffer from a reduction in the access to and the range of local bio-resources that are used for their daily lives. Men and women, local business men and groups from neighbouring but different ecological zones will experience reductions in biodiversity in different ways (and the gender dimension is discussed in another chapter by Jackson in this volume).
- International firms in the business of biotechnology will be interested in prospecting a very wide range of genetic resources, and in maintaining access to the rights to research the almost limitless opportunities in landraces and in tropical forests.

Both these actors define biodiversity (and its reduction) in different ways and at different levels of aggregation, and yet they are in competition for many of the same bio-resources. Furthermore, the ways in which biodiversity is invoked and measured are diverse, as Stocking et al. in this volume explore. It is very easy to use statistical means (by limiting focus on short-term, taxonomically narrow and spatially limited considerations) to lie about biodiversity for a variety of political and professional ends. Statistics are socially constructed too, even those compiled with honest intentions.

Another example of a highly socially constructed "scientific" issue is that of sustainability, discussed in this volume by Gibbon et al. Their chapter consists of a discussion of unresolved (and perhaps unresolvable) debates about what the term means and what its scientific content should consist of. At the other end of the continuum is the example of acid rain. The symptoms of the problem were recognized in Europe in specific countries, and the scientific work on tracing causes was carried out by the most affected countries, whereas others that were identified as major contributors (e.g. the United Kingdom) dragged their feet on initiating their own research. However, the causes have now been established beyond reasonable doubt, even beyond that employed as an excuse for inaction and waiting for further confirmation. As science has established with increasing clarity the processes involved in the creation of acid rain, alternate explanations and political prevarications became impossible. The contribution of industrial pollution to acid rain became uncontroversial, and accepted as "real".

10

In these latter and simpler cases, perceptions and problem definitions are widely shared. Normal "objective" science, at least for relatively tame and carefully bounded problem-solving areas, will continue to play a central role in understanding society–environment relationships in a conventionally defined role. This will be so even at the interface with a critical social science in political ecology. On the other hand, the subject matter of most natural science research is embedded in notions that are more obviously socially constructed, because they are culturally embedded. For example, the issues of degradation, pollution, or risk may imply concepts for natural scientists, which are analyzable by normal positivist natural science, and for whom the only problems are technical ones (e.g. of definition, data collection and experimental design). However, all these issues are pregnant with social significance and are subject to a rich variety of interpretations (see Douglas & Wildavsky 1982 for a social analysis of risk). They simply cannot be captured by a single and authoritative scientific set of facts. In all these cases, their natural science components are imbued with judgments about:

- scientific agenda (what gets studied and what ignored)
- how scientists study them (the institutions in which they study, and the reward structures for doing so)
- the ideological assumptions about the terms themselves (sustainability, degradation or biodiversity for whom, under what social relations, with costs and benefits for whom, and soon).

In these cases, the scientific problem area is anything but tame, and scientific problem definition, method and interpretation are (like all science) partisan, but in such cases especially so.

It may sound gratuitous to defend a role for objectivist natural sciences in this volume, which is written mostly by natural scientists, but a long-established critique of science from a variety of ideological standpoints (liberal, neo-Marxist, Frankfurt School and postmodern) has tended to undermine altogether the claims of legitimacy and authority of much environmental research. Bloor (1976), Latour & Woolgar (1979) and more recently Jasanoff (1990) have examined the ways in which scientists actually work, and show that even the "hardest" science and the most "objective" method of scientific replication are negotiated between scientists and are thoroughly socially constructed.

However, a counterweight to the deconstruction of science must also be provided. A case could be made that the bulk of what is styled as political ecology has been written by social scientists, who have paid little attention to what natural scientists have had to say about their environments, usually with embarrassing results. One example of an ideological offensive unfettered by any understanding of the actual nature of environmental change, which might have been provided by natural scientists, is the radical critique of desertification in the 1970s. This debate unproblematically attributed desertification to the deepening of commodity relations in the periphery of a world capitalist system. A recent letter to the *New York Times* (March 1994) outlines the findings of Hellden (1991) and other Swedish researchers (Jirstrom & Rundquist 1992, see also Nelson 1988 for earlier critical work) on the causes of desertification, and who attribute the major cause of vegetation loss and localized invasion of wind-blown sand to climatic fluctuations, and attested to the resilience of the ecosystem after the resumption of more normal patterns of precipitation. So, the case for the globalization of capital being causal in desertification looks rather amateur, since the scientific evidence of permanent damage to the environment points in other directions. However, its causal role in the region's indebtedness, immiseration and starvation of the peoples of Sahelian Africa remains an arguable one, but it will remain so on the grounds of (unfalsifiable) ideological force of persuasion. For want of attention to a large and accessible body of climatological and ecological information, the case for adding desertification to the long list of other socially induced woes, now looks very thin.

The central elements of political ecology

Political ecology has recently used a dialectical approach both to the mode of discourse of social and environmental relations as well as to the explanation of those relations themselves. As Howard & King (1975: 21) define the dialectical process after Marx:

> . . . the main idea is that social phenomena are seen as existing in relation to each other, and continually developing in and through such relations so as to form at various phases contradictory forces that generate qualitatively new formations. Thus, social reality is always seen as in a state of becoming something else.

In the light of the dialectical approach, the central ground of political ecology may be stated thus:

> Environment can be seen as "enabling", or "affording" in the sense of providing resources and services as they are defined and redefined by society as it develops. Environment therefore is constantly in a state of being conceived of, learnt about, acted upon, created and recreated and modified, thus providing a constantly shifting "action space", both productive and ideational for different players, as they create and

recreate their own history. At each moment in these histories then, the environment is in a reflexive relation to these different players in which it offers both opportunity and constraints. These are both socially patterned through access, use and control of elements in the environment; and environmentally patterned by physical limits, which themselves are subject to available and differentiated knowledges, technologies, labour and capital.

Let us unpack this statement of the field of political ecology, and examine its components:

- Political ecology must use the epistemologies of its parent disciplines. These obviously come from both the social and the natural sciences and will be different, mainly regarding the status and domain of proof. More specifically, the practice of "normal" science as the sole vehicle of an objective, reductionist and universal truth about environment and society has to be renegotiated. This issue therefore becomes one of the central epistemological grounds for political ecology, and has already been introduced above. In practice this is seldom easy, since each will be required to accept quite fundamental critiques from other epistemic communities, which are often institutionally formed in competitive and mutually antagonistic ways. For all the rhetoric and examples of creative interdisciplinary work and institutional forms (see Checkland 1985, Checkland & Scholes 1990), there still remains a good deal of distrust and disrespect between natural and social scientists.

- Natural scientists do not like to be told that even the "hardest" objective scientific methodologies (e.g. replication) are socially constructed and open to social negotiation. There are undoubted normative and moral considerations that are implied in logical positivistic science, which may be perceived to be called into question by social critique.

- Equally, social scientists may not take kindly to the charge of amateurishness and ignorance about the (environmental) object of study. Simply put, they require a grasp of the principles of natural science, without which their contribution is all critique and no substance about the interactions and outcomes within the natural world.

- The complex interactions between environment and society are always embedded in history and locally specific ecologies. Therefore, it is likely that long historical periods of time have to be considered. Although "a day may be a long time in politics" (an aphorism coined by Harold Wilson, the British Prime Minister in the 1960s), a decade or even a century may be too short a perspective for these interactions to work through as environmental change and its social impacts (the case of global warming is a case in point). Land degradation, pollution, reductions in biodiversity, and global warming may take many decades. Thus, in many political ecologies a view of the long term may afford the only credible one.

An example of the richness (and scientific reliability) of such a long-term approach can be found in Tiffen & Mortimore's (1993) analysis of the Machakos region in Kenya from the 1920s to the present. This study emphasizes the reproduction of both ideas about, and the objective conditions in the physical environment. Nibbering's (1992) two-millennium study of an area in southwest Java, and that of Hefner (1990), also in Java, are other tours-de-force with obvious advantages in the long view.

Another challenge that emerges from studies of the longer term is that landscapes and natural systems may seem to have changed less than the views of them. For example, one of the unsettling conclusions of Tiffen & Mortimore's study is that the case of environmental degradation has been made several times before by outsiders and has proved inconclusive. What should we be examining now that possible new problems in Machakos are appearing: the landscape and its people in a contemporary context once again, or our own optic through which we have been viewing them?

- Since political ecology is usually based, as part of its concerns, with place-based and locally specific interactions and also in larger, pervasive and often non-place-based political and ideational forces (e.g. environmental ideologies, State policies), several different levels or scales of analysis are implied. These have to be made specific (a relatively easy task) and linked by credible explanations. The latter is very much more problematic, involving the internalization and linking of proximate and remote causes.

- Many discourses about the environment are conceived of at a global scale, and are also pursued internationally (Dalby 1992). They find voice in small but powerful groups and networks in universities, aid agencies and international institutes. Leading opinion-formers within these select institutions may react upon each other to form broad agreement and shape the course of future debate (and research funds). These are examples of what Ruggie (1975) calls "epistemic communities".

International institutions can pack formidable knowledge claims and transmit them to individual States and civil societies throughout the world. They can achieve this through economic conditionality of multilateral and bilateral loans, local bargaining over access and exploitation of natural resources, and their claims to scientific rationality and expertise.

Although structuralist explanations have a certain force in explaining their content and why they appear on agendas at particular times, other explanations have to be brought to bear concerning how scientists and international bureaucracies and organizations work from day to day, and how they deal with contest and dissent. Thus, a critical analysis of international policy about the global environment, together with its politics, must remain an important concern.

The State, as conceived of as an idea, and also its institutions (legislature, departments of agriculture, right down to its street-level bureaucracies) must be central to political ecology. The role of the State in environmental management in developing countries is also a major current debate. The World Bank, for example, has established a neo-liberal view, based upon the new institutional economics, which assumes a very different role from earlier more interventionist ideas (for a critique see Biot et al. 1994 forthcoming).

Empirical support of the importance of the State in environmental policy and management if it is needed, can be found in a north–south traverse across the countries of central Africa (the Sudan, Kenya, Tanzania, Zimbabwe and South Africa). This reveals an extraordinary variety in the nature of these States, in their environmental policies, land tenure and their abilities to intervene in resource management within their territorial borders. Forests provide one of the most important loci of the political struggle over resources. Examples of excellent case studies on the role of the State in conserving the forest against local and poor people can be found in Hecht & Cockburn (1989) in the Amazon, and in Peluso (1992) in Java.

Conflict and contestation over the environment are central concerns of political ecology. These may take both direct forms leading to struggles over "who gets what" (e.g. Moore 1993 in which "contesting terrain" in Zimbabwe is examined using a political ecology approach), but also as ideas that do not explicitly address struggles of unadorned economic interest at all. Examples of the latter lie more in ideas and meaning (e.g. biodiversity, sustainability, soil and water conservation, and public campaigns such as "save the rainforest"). In these cases discourse can less readily be deconstructed by a simple identification of economic interests (Dalby 1992, Taylor & Buttel 1992). However, both types of contestation are central to the politics of political ecology.

A specific form of these contestations has been between knowledge systems about the environment, particularly between exogenous (Western) and indigenous knowledge (IK). There is an immense literature on this subject,

(Chambers et al. 1989, Bebbington 1991, Fairhead 1991, Rajasekaran et al. 1991, Dissanayake 1991, Cornwall et al. 1992, Cromwell et al. 1992, Pretty & Chambers 1993). Dominant paradigms of western scientific thought have coexisted with and overlain an array of local knowledge systems. Part of the growing realization that "modern" and "sustainable" technologies have caused environmental damage and immiseration has prompted a new dialogue with IK's. Part of this discourse is congruent with neo-populist approaches to development, but other approaches (the classic and neo-liberal) view the importance of IK in a very different light. These views regarding the environment, its managers and their knowledge are discussed below.

There are two major sets of reasons why political ecology has also focused on the micro-level. The first is that the immediate users of resources and/or proximate causal agents of environmental change are the first link in any explanatory model in political ecology. Thus, the industrial manager, the directors of a logging company, the farmer and pastoralist will be centre-stage. Of course it is not straightforward to identify who the decision-maker may be. Is it the person who does most of the agricultural work in developing countries (the woman), or the person who directs operations, maybe in an indirect and delegatory manner (the, usually male, "head of household")? Secondly, how decisions about resource use and management are made is also one of the keys to understanding, particularly for the policy-maker. Theories of individual and collective action abound from economics, sociology and organization theory, and they form coherent ideological statements about the nature of society and (specifically) environmental management. It has been common to contrast economists' understanding of human action, as the result of decision-making under constraints, with that of sociologists who see action as determined by structure. It is said that economists see nothing but choice, whereas sociologists see no possible role for choice. This distinction may be overdrawn, but it highlights the interdisciplinary requirements for research, where political ecology may have a comparative advantage.

Recent developments in sociology and anthropology have helped to "free up" the historical tensions in the social sciences between macro and micro, between structure and agency, and between determinism and voluntarism. The idea of agency (after Giddens 1979) is particularly liberating, where the actions of agents differs from the individual decision-maker, in that the actions of agents are socially structured, rather than determined by rational choice. Individual action is seen to have intended and unintended outcomes and is formed by reflective monitoring of action and its rationalization, together with unconscious motivations that can be accounted for by social structure and socialization.

16

Zimmerer (1991) has used these theoretical advances to good effect in a study of changing social relations of production, peasant resistance and the draining of montane wetlands in highland Peru. Further and parallel perspectives of Long & Long (1992) take an actor-orientated approach, in which individuals' actions may relate to internal negotiations between different identities, which will be recognized by an individual. Thus, someone may be a mother, a wife, a trader, a fuel gatherer, and a grower of household staples. How she will act will be an outcome of different sets of negotiations and trade-offs between her various identities and "projects", which involve both introspection and responses to others (e.g. struggle, evasion, compliance). It is at this level that an innovative political ecology should provide many opportunities for theoretical and empirical work.

These elements have been discussed as an agenda for a political ecology in the 1990s, and their identification locates the ground of this developing field. The point is that they have been discussed sequentially, but are dealt with and used simultaneously in any study of political ecology. It is the disciplinary, epistemological and methodological contradictions that these elements bring together that gives it such vitality. Thus, rather than defining political ecology (which focuses on limits and where the stimulus for creative work is growing weak or non-existent), it is perhaps better to identify the areas of association and contradiction within and between these components (and perhaps other candidates not mentioned here). The more the subject matter may contain these elements, the closer it will be to the core of a coherent political ecology.

In the second half of this paper, these components are illustrated initially as a general case, as they are organized into a hierarchy of scales, and then illustrated in more detail taking the case of land degradation.

The chain of explanation – the general case

Political ecology usually combines an analysis of "place-based" and local elements, both physical and human, with that of other "non-place-based" elements that are usually ideational, and is often pitched at several different levels. Blaikie (1989) developed "a chain of explanation" for the specific case of land degradation, building upon the original idea in Blaikie & Brookfield (1987), which has been critiqued (Black 1990, Peet & Watts 1993) and modified since. Figure 1.1 gives a schematic hierarchy of levels (and where these are given spatial expression, scales), and gives some examples of the main links under study in three examples.

17

SITE↔SYMPTOM↔PRACTICE↔"DECISION–MAKING"↔SOCIETY↔STATE↔WORLD

Figure 1.1 A chain of explanation for political ecology

In this schema the site of environmental change (whatever aspect may be chosen and recognized, be it industrial or marine pollution, land degradation or other environmental changes) is the starting point of the chain of explanation. In order for these changes to become an object of thought, they will have to be recognized as having an impact in some way upon (certain elements in) society. As indicated earlier in this chapter, the way in which information about the state of the environment is produced, controlled and censored should be considered at this point. This is particularly so in slow-acting and invisible impacts of environmental change, as in ozone depletion, acid rain, man-made nuclear radiation and many other forms of pollution. These impacts may be deleterious or advantageous for different people at different places, and may have varying impacts upon them. These impacts may be styled as economic symptoms of environmental change, and have been caused by specific productive activities that usually have to be identified. These activities may be characterized by the detail of the technologies of production most linked to the environmental change in question (from industrial solid waste disposal, scrubbers for airborne factory wastes, through to the design of trawler nets, or bench terraces for paddy agriculture).

In turn, these technologies are chosen and operated by specific agents, sometimes characterized as decision-makers in the context of their immediate decision-making environment. These agents, of course, are part of a wider civil society, and will be part of a political economy, and be linked to it through the social relations of production and a culture that structures this decision-making environment.

At the next link in the explanation, the State will usually have subtle and profound effects upon both civil society and the specific agents immediately instrumental in making productive decisions, which may lead to degradation. At the level of the State, national politics, often using "science" (often selectively) as a source of authoritative knowledge, will provide a room for manoeuvre for the agents in question.

Finally, nation States are part of the global economy and will be subject to the dynamic of the globalization of capital through such global institutions as commodity and financial markets, debt relations, foreign aid and conditionality of loans. It must be emphasized that this chain stretches the length of political ecology, but any one empirical or theoretical study will usually involve one or more of these links, as the examples in Figure 1.1

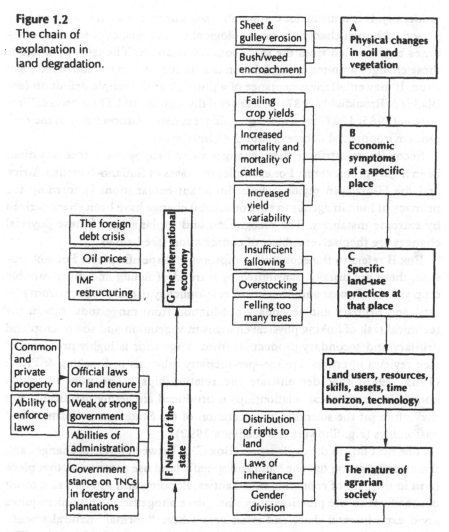

Figure 1.2
The chain of explanation in land degradation.

show. Also, the geometry of these links and the direction and type of explanatory link may also be altered to focus on the issue at hand. Some levels may be inappropriate and omitted altogether.

The easiest task is to identify the links, but it is more difficult to operationalize them. As an illustration of some of these difficulties, an example will be taken of the chain of explanation in land degradation (see Fig. 1.2). In this configuration, environmental change has been made the starting point (Box A). To identify physical changes in soil and vegetation at a specific place usually requires good time-series data. Satellite imagery, aerial photography, travellers' records and oral testimony (usually of the aged) are typical sources. However, physical changes in soil are usually more difficult to identify. It is unusual for comparable soil samples to be taken over a long period, although changes in morphological characteristics of soil can sometimes be identified from the same sources as above. The issue of whether these changes amount to degradation is a difficult technical and ideological issue. It may entail the acceptance of a pluralist and multiple definition (see Blaikie & Brookfield (1987) for a general discussion, and Thompson & Warburton (1985, 1987) for a more specific treatment of uncertainty in the current environmental debates about the Himalayas).

Secondly, to attribute these changes to anthropogenic forces has often been highly problematic. For example, the cases of Sudano–Sahelian Africa and the Himalayan region, where dominant explanations concerning the primacy of human agency in environmental change have been characterized by extreme instability. The recognition and explanation of these physical changes are themselves subject of contestation (see below).

Box B refers to the economic symptoms at a specific place. For soil erosion, these are usually transmitted by a trend of falling and more variable crop yields, biomass and diversity of economically useful products from forests, and primary and secondary production from rangelands. Again, the technical task of linking physical changes in vegetation and soil to crop and primary and secondary production from rangelands is highly problematic (see reviews of crucial erosion–productivity relations in Stocking & Peake (1985, 1986) that demonstrate the relationships). The ways in which knowledge about these relationships is produced are highly significant, and they often pit the scientific interpretation of the State against farmers and pastoralists (e.g. Showers & Malahleha 1992).

The next link in the explanation (Box C) is between physical changes and their symptoms on the one hand and specific land use practices at that place (and in the case of economic externalities, flooding and siltation as a result of specific land-use practices in another place altogether). This link requires good experimental data, and is an area where "normal" natural science (hydrology, geomorphology and soil science) has a central role to play. The enormous volume of literature, often highly disputational, is witness to the levels of uncertainty in making this link outside the research station and the soil erosion plot (Blaikie & Brookfield 1987). Again, the literature on the Himalayas, the desertification debate, and overstocking controversies in sub-Saharan Africa demonstrates how much this link is open to uncertainty, both between scientists and between them and local resource users.

The next levels of explanation (Boxes D and E) are usually pivotal and the most complex. This concerns the ways in which individual and collective action and technical choice in the use of natural resources leads to the specific land-use practices to be explained. Access, use and control of those natural resources are usually central in explaining how they are used, and this is socially structured at the individual and "household" level. The gender division of labour within households can cause specific labour bottlenecks and also bring to bear a gender-specific knowledge in the use of those natural resources. There are also other mechanisms that structure access to resources such as age, seniority, kinship and clan. Struggles over access to resources and outcomes, in terms of their management, take place in several different arenas:

- internal negotiations between different identities within an individual
- between men, women, seniors and juniors
- within a household and between households (or their, usually male, representatives)
- between the above within local institutions
- between interest groups (and more broadly, classes) within the regional political economy.

For an example of a "stakeholder" approach to a national park in Zambia, see Abel & Blaikie (1986). Examples of gender-centred struggles for control over new resources in Gambia can be found in both Carney (1993) and Schroeder (1993). The former example is in relation to irrigated wetlands, whereas the latter is in relation to orchards and gardens.

A major focus of attention in resource management and conservation has been local institutions for the management of common property. A thematic discussion of this focus along with case studies can be found in Ostrom (1990) and Bromley (1992). Abel & Blaikie (1988) provide a comparison of the social and technical aspects of communal range management in Zimbabwe and Botswana. The issue of property regimes and their capabilities for sustainable management spans local perceptions and struggles (and therefore local politics) and the influence of the State. Contradictions between customary and official land law throughout most of Africa is a case in point (see examples in Bassett & Crummey 1993). The erosion of local authority (usually of a chiefly or feudal kind) by the penetration of forms of peripheral capital into rural areas of lesser developed countries has been an enduring theme (see case studies in BOSTID 1985, and Bromley 1992).

The State (Box F) has long played a commanding role in environmental management of its territory, although often with unintended and sometimes disastrous results. Notions die hard of the Weberian State that rationally intervenes based on objective scientific information above competing interests to conserve the environment for future generations. The notion dies hard that the State rationally intervenes in order to restore sound environmental management. Critiques of this (still dominant) view maybe is still one of the central concerns of a critical political ecology. In the case of land degradation, the links have been well documented between the various and sometimes contradictory interests of the colonial State (protectorate, settler State or other) and soil and water conservation policy. However, the links between so-called "interests" and State conservation policy are much less strongly and simply determined by economic ones alone. More detailed study of how scientific information about land degradation is actually produced, and of how scientists actually work, is usually required, and the outcomes in terms of what the State eventually calls policy are often highly contingent.

A brief example from sub-Saharan Africa of how "the [scientific] institutions are the facts" (the aphorism is from Thompson & Warburton 1987) can be taken from recent reassessments of range ecology (Abel & Blaikie 1988, Benhke & Scoones 1992, Scoones 1992). The concept of stocking density, which was a major planning tool in sub-Saharan Africa for over 50 years, was developed, and it used supporting experimental data from research stations. It was believed that a theoretical maximum density (expressed in hectares per livestock unit) for ranges of different soils and precipitation could be calculated, which would not "degrade" the range. The technical models of what constituted degradation and "optimum" carrying capacity are excluded here for the sake of brevity, but see Behnke & Scoones (op. cit.: ch. 6) for a review. The experimental data, on which many of the conventional views about range degradation were based until very recently, were derived from a series of experiments in which cattle of different densities were pastured on ranges of different qualities. The vegetation changes and cattle quality were observed to provide these theoretical maxima (or carrying capacity). However, the actual practice of pastoralists differed from that reproduced from research station conditions in two main respects:

- First, pastoralists follow an opportunistic herding strategy in which the cattle are taken to where recent precipitation and current forage exist. Densities of cattle are therefore very high in particular localities, but for short periods. Expertise is also used in moving cattle across the range to conserve the range itself and to provide maximum access to forage for the cattle. This expertise could not be reproduced under experimental conditions.
- Secondly, cattle condition (which is the major determinant of price when the cattle are sold for meat) may not be a major consideration for African pastoralists. Annual off-take for sale in pastoralist herds is very much lower than for commercial herds, since commercial production is not the major objective. Therefore, large number of cattle (but in poorer condition) are preferred as an insurance against

drought and raised cattle mortality which, as pastoralists knew, was bound to occur from time to time.

Thus, it can be seen how accurate and costly research information was produced by research stations in the colonial and post-colonial State, which, although validly linking stocking density to range condition, did not include the objectives and practice of pastoralists within the experimental design. The charge is persuasive that the authoritative and scientific knowledge derived in this way was used to justify the encroachment and privatization of the range by commercial ranchers. However, this is unlikely to mean that scientists actively colluded with settler interests, nor that they falsified their results to please their political masters (although there are cases of this too). Their research and the institutions that produced it are linked in a much more contingent and complex manner than a simple structuralist explanation would have us believe.

Global processes and global discourses

The final level of analysis, the least proximate to our place-based starting point, is the international economy (Box G). At this level, the process of the globalization of capital will be a central starting point. Its environmental impacts, such as land degradation, can be analyzed at a variety of levels (local, watershed, national and global). The global level tends to involve more generalized calls for a new economic order, and a transformation of North–South relations. These solutions can be of a radical nature and they involve the nature of (capitalist) growth itself. The *Limits to growth* thesis (Meadows 1972, 1992) is another major non-radical technical attempt at the global level. Land degradation on a global scale may have significant impacts upon global warming through the removal of forests and other vegetation. Also, changes in habitat may threaten biodiversity.

However, a more common concern is how the global economic system impacts upon States through the universalization and export of neo-liberal economic doctrines, and through the active penetration on the ground of commercial agriculture. One such research area is the supposed environmental impact of structural adjustment policies developed by the IMF/World Bank. Reed (1990) has edited a volume on the supposed impacts, and mention is made of the problems of proof of assertions, one way or the other. Mearns (1991) started a similar project to link structural adjustment to environment in Malawi, but ended up writing a much more interesting piece on scientific method. One of the main points he made was that it was

23

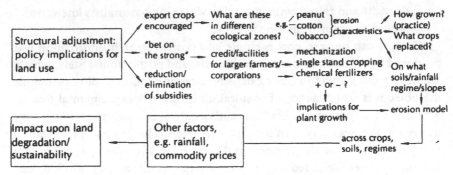

Figure 1.3 Modelling the impact of structural adjustment.

very difficult to achieve proof on the matter using the tools of "normal" science.

Figure 1.3 shows the necessary research area in one small part of the linkage between the supposed independent variable (structural adjustment) and dependent variable (environmental degradation). There are formidable difficulties in creating a credible experimental design in the face of critical data shortages, *non ceteris paribus* (e.g. 500 000 Mozambiquan refugees, drought, moves in relative commodity prices during the period under study). It is no wonder Peet & Watts (1993: 239) found Blaikie & Brookfield's explanatory linkages to be "woolly". It is, as Mearns put it, a genuinely "wicked problem area" and one that will exercise the ingenuity of political ecologists that demands a switch of scales and epistomologies simultaneously! A preliminary discussion of this area and a recommended research approach called "progressive contextualization" is provided by Vayda (1983).

A dominant mode of discourse about the global environment contains notions concerning global environmental management and the global commons. Such institutions as the World Bank, the International Union for the Conservation of Nature (IUCN), the World Wildlife Fund (WWF) and the International Whaling Commission (IWC) are responsible for promoting their environmental agendas on a world scale. Also the UNCED meeting in Rio provided a spectacle of naked economic and strategic interest, political priorities and a wide array of non-governmental organizations, pressure-groups and token oppressed minorities all submitting to international bureaucratic dictat (Grubb et al. 1993).

There are also several major international environmental accords that claim to facilitate global environmental management, as has already been mentioned earlier in this chapter. Whether they are ratified, implemented and complied with is an important and interesting question. At present there is a multinational research program entitled Strengthening Compliance with International Environmental Accords, financed through the SSRC (and see also Haas et al. 1993). It is mainly pursued through the tools of analysis of international law and politics. However, such local studies as there are (of compliance at the level of the State-civil society interface, and of local decision-making) employ a political ecology approach, involving almost all the elements discussed above.

Lastly, there are global environmental approaches made by international institutions, which amount to articles of faith, buttressed by recourse to theory. These form a central part of global environmental discourses, in which much natural and social scientific research is reinterpreted and incorporated. Each of the three paradigms are (more or less) internally consistent, according to several diagnostic variables (of which some are expanded into whole theories). There are presently three sets of assumptions on which issues of land degradation in developing countries rest:

- The first set assumes that the extent of and solutions to the problems of land degradation are well known, but the problem is to get people to implement them.
- The second set assumes that the nature and extent of land degradation are imperfectly understood, that local peoples often reject conservation technologies for good reasons, and in fact adopt their own individual and collective approaches that have in the past resulted in sustainable livelihood practices (see IFAD 1992 for an annotated guide and commentary for such techniques in sub-Saharan Africa).

Norman Uphoff has characterized these contrasting positions (with respect to poverty alleviation policies and projects) as "paternalist" and "populist" respectively. The former is characteristic of soil and water conservation (SWC) in the colonial and ex-colonial territories of East, Central and Southern Africa, and South Asia and of the American Soil and Water Conservation tradition. However, it has regained favour, although in a new guise, as a part of the third set of assumptions.

- The third set is a new synthesis of previous views, which may be called the "neo-liberal" approach. This approach does not see the availability of technologies as problematic. Instead, it is the provision of incentives to conserve land that should be emphasized. This involves the evolution of appropriate property rights, the scrapping of "perverse" incentives that encourage degradation, and the stepping back by the State to a role of "referee" for the efficient operation of markets (see Biot et al. 1995 for an expansion of these themes).

Although no institution will follow one school of thought in its pure form (institutions are seldom monolithic), it will usually privilege one at their expense of others. Sometimes others will be allowed to exist in a marginal

25

form as a way of incorporating them and subjugating them to the hegemonic under the name of free speech and "let a hundred flowers bloom". Political ecology can play a role in deconstructing these views and how they are used in discourse and in policy formulation. However, its comparative advantage will lie in extending the discourse to empirical and often local research.

Conclusion

Environmental discourses imply that experience and interpretation of environmental change are varied and contested. Information about these changes derive equally from a great variety of sources. Although information from natural scientists may often be claimed as authoritative, it is produced and used selectively. Therefore, it can never be understood solely from the perspective of the natural sciences. Scientists are but one set of players in the experience and interpretation of the environment. They also negotiate, dispute and propagate their own ideas, with others.

In this chapter, political ecology is presented as a means of understanding the environment as both socially constructed and partly understandable through scientific measurement and interpretation of physical processes. Political ecology can locate itself at points of new association and contradiction from its parent disciplines in the natural and social sciences. It should identify what these points are, and this paper has suggested several, but there will be others that could probably press their case. Then, those points should be linked as markers for a methodology and empirical field. They are bound to be problematic and challenging, which will stimulate creative new thinking rather than a drive for disciplinary hegemony and academic territory. New thinking about the environment can be achieved only by sharing an agenda that confronts different political and professional approaches. This is best done by sharing it in practice: in study, in activist and advocacy roles and in policy. It is much easier to demonstrate swimming by doing it than by standing by the pool explaining how it should (or should not) be done. The ensuing chapters in this book should help show the way.

Bibliography

Note: there are twenty references included in this bibliography as further useful reading, but are not referred to in the text.

Abel, N. O. J., & P. M. Blaikie 1986. Elephants, people, parks and development: the case of the Luangwa Valley, Zambia. *Environmental Management* 10, 735–51.

— 1988. *Managing common property resources in rural development: the case of Zimbabwe and Botswana.* Mimeo, Overseas Development Administration, London.

Alverson, H. 1984. The wisdom of tradition in the development of dry-land farming: Botswana. *Human Organisation* 43, 1–8.

Bassett, T. J. 1988. The political ecology of peasant–herder conflicts in the northern Ivory Coast. *Association of American Geographers, Annals* 78, 453–72.

— 1993. Political ecology and ecological theory: towards a critical integration. Paper presented at Africa Studies Association conference, December 4–8, Boston.

Bassett, T. J. & D. E. Crummey (eds) 1993. *Land in African agrarian systems.* Madison: University of Wisconsin Press.

Bebbington, A. 1993. Modernization from below: an alternative indigenous development. *Economic Geography* 69, 274–92.

Behnke, R. H. 1994. Natural resource management in pastoral Africa. *Development Policy Review* 12, 5–27.

Behnke, R. H. & I. Scoones 1992. *Rethinking range ecology: implications for rangeland management in Africa.* London: Dryland Networks Programme.

Biggs, S. D. & J. Farrington 1990. *Assessing the effects on farming systems research: time for the reintroduction of a political & institutional perspective.* Paper prepared for the Asian Farming Systems Research & Extension Symposium, 19–22 November, Bangkok, Thailand.

Biot, Y., P. M. Blaikie, C. Jackson, & R. Palmer-Jones 1995. *Rethinking research on land degradation in developing countries.* Discussion Paper 289, World Bank, Washington DC.

Black, R. 1990. Regional political economy in theory & practice: a case study from Northern Portugal. *Institute of British Geographers, Transactions* 15, 35–47.

Blaikie, P. M. 1985. *The Political Economy of Soil Erosion in Developing Countries.* London: Longman.

— 1989. Explanation and policy in land degradation and rehabilitation for developing countries. *Land Degradation and Rehabilitation* 1, 23–38.

Blaikie, P. M. & H. C. Brookfield 1987. *Land degradation and society.* London: Routledge.

Bloor, D. 1976. *Knowledge & social imagery.* Chicago: University of Chicago Press.

BOSTID 1985. *Common property resource management* (proceedings of a conference: Panel on Common Property Resource Management). Washington DC: Board of Science & Technology of International Development, Natural Resource Council.

Bromley, D. W. 1992. *Making the commons work.* San Francisco: ICS Press.

Bryant, R. I. 1992. Political ecology: an emerging research agenda in Third World Studies. *Political Geography* 1, 14–36.

Carney, J. 1993. Converting the wetlands, engendering the environment: the intersection of gender with agrarian change in Gambia. *Economic Geography* 69, 329–48.

Chambers, R., A. Pacey, L. A. Thrupp (eds) 1989. *Farmer first: farmer innovation and agricultural research*. London: IT Publications.

Checkland, P. 1985. From optimising to learning: a development of systems thinking for the 1990s. *Journal of Operational Research Society* **36**, 9.

—& J. Scholes 1990. *Soft systems methodology in action*. Chichester: John Wiley.

Cockburn, A. & J. Ridgeway (eds) 1979. *Political ecology*. New York: Times Books.

Colchester, M. 1993. Slave and enclave: towards a political ecology of equatorial Africa. Paper presented at the IWGIA Conference on The Question of Indigenous Peoples in Africa, Copenhagen, 1–3 June.

Cornwall, A., I. Guijt, A. Welbourn 1992. Acknowledging process: challenges for agricultural research and extension methodology (Overview Paper II). IIED/IDS workshop (27–9 October), IDS, Brighton.

Dalby, S. 1992. Ecopolitical discourse: environmental security and political geography. *Progress in Human Geography* **16**, 503–522.

De Janvry, A. & R. Garcia 1992. *Rural poverty and environmental degradation in Latin America*. Staff Working Paper 1, IFAD, Rome.

Dissanayake, W. 1991. Knowledge, culture and power: some theoretical issues related to the agricultural knowledge and research system framework. Agric. Knowledge Systems & the Role of Extension. *Proceedings of International Workshop*, Bad Boll, 21–4 May 1991.

Douglas, M. & A. Wildavsky 1982. *Risk and culture: an essay on the selection of technical and environmental dangers*. Berkeley: California University Press.

Dove, M. R. 1984. *Government versus peasant beliefs concerning* Imperata *and* Eupatorium: *a structural analysis of knowledge, myth and agricultural ecology in Indonesia*. Mimeo, East–West Center, University of Hawaii, Honolulu.

Drinkwater, M. 1991. *The state & African change in Zimbabwe's communal areas*. London: Macmillan.

Eckersley, R. 1992. *Environmentalism and political theory: towards an ecocentric approach*. London: UCL Press.

Ensenberger, H. M. 1974. A critique of political ecology. *New Left Review* **8**, 3–32

Essel, J., & R. Peet 1989. Resource management and natural hazards. In *New models in geography: the political economy perspective*, R. Peet & N. Thrift (eds), 49–76. London: Unwin Hyman.

Fairhead, J., 1991. *Indigenous technical knowledge and natural resource management in sub-Saharan Africa: a critical review*. Paper commissioned by SSRC Africa Agriculture Project.

Fosbrooke, H. & R. Young 1976. *Land and politics among the Luguru of Tanganyika*. London: Routledge & Kegan Paul.

Giddens, A. 1984. *The constitution of society: outline of the theory of structuration*. Cambridge: Polity.

Gritzner, J. 1988. *The West Africa Sahel: human agency & environmental change*. Geography Research Paper 226, University of Chicago.

Grubb, M. Koch, M. Munson, A. Sullivan, F., K. Thomson 1993. *The Earth Summit Agreements: a guide and assessment*. London: Earthscan.

Haas, P. M., R. O. Keohane & M. A. Levy 1993. *Institutions for the Earth: sources of effective international environmental protection*. Cambridge: MIT Press.

Hecht, S. & A. Cockburn 1989. *The fate of the forest: developers, destroyers and defenders of the Amazon*. London: Verso.

Hefner, R. W. 1990. *The political economy of mountain Java: an interpretive history*. Berkeley: University of California Press.

Hellden, U. 1991. Desertification – time for an assessment. *Ambio* 20, 372–83.

Howard, M. C. & J. E. King 1981. *The political economy of Marx*. London: Longman.

Jasanoff, S. 1990. *The fifth branch: science advisers as policy-makers*. Cambridge, Mass.: Harvard University Press.

Jirström, M. & F. M. Rundquist 1992. *Physical, social and economic aspects of environmental degradation*. Rapporter och Notiser 108, Institutionen för Kulturgeografi och Ekonomisk Geografi, University of Lund, Sweden.

Kuhn, T. S. 1962. *The structure of scientific revolutions*. Chicago: University of Chicago Press.

Latour, B. & S. Woolgar 1979. *Laboratory life: the construction of scientific facts*. Princeton: Princeton University Press.

Long, N. & A. Long (eds) 1992. *Battlefields of knowledge*. London: Routledge.

Mazur, R. E. & T. Titilola 1992. Social and economic dimensions of local knowledge systems in African sustainable agriculture. *Sociologia Ruralis* XXXII 264–86.

Mearns, R. 1991. *Environmental implications of structural adjustment: reflections on scientific method*. Discussion Paper 284, IDS, Brighton.

Meadows, D. H. 1992. *Beyond the limits: global collapse or a sustainable future*. London: Earthscan.

Moore, D. S. 1993. Contesting terrain in Zimbabwe's Eastern Highlands: political ecology, ethnography and peasant resource struggles. *Economic Geography* 69, 380–401.

Nelson, R. 1988. *Dryland management: the desertification problem*. Environmental Department Working Paper 8, World Bank, Washington DC.

Nibbering, J. W. S. M. 1991. *Hoeing in the hills: stress and resilience in an upland farming system in Java*. PhD thesis, Research School of Pacific Studies, Australian National University.

O'Riordan, T. 1981. *Environmentalism*. London: Pion.

Ostrom, E. 1990. *Governing the commons: the evolution of institutions for collective action*. Cambridge: Cambridge University Press.

Peet, R. & N. Thrift 1989. Political economy and human geography. In *New Models in Geography*, R. Peet & N. Thrift (eds), 3–29. London: Unwin Hyman.

Peet, R. & M. Watts 1993. Introduction: development theory and environment in an age of market triumphalism. *Economic Geography* 69, 227–53.

Peluso, N. L. 1992. *Rich forests, poor people: resource control in resistance in Java*. Berkeley: University of California Press.

Pepper, D. 1984. *The roots of modern environmentalism*. London: Routledge.

Porter, D. 1993. The World Bank and the environment: a few green threads in *The Emperor's* clothes. Mimeo, Research School of Pacific Studies, Australian National University, Canberra.

Pretty, J. N. & R. Chambers 1993. *Towards a learning paradigm: new professionalism and institutions for agriculture*. Discussion Paper 334, IDS, Brighton.

Pryce, R. 1977. Approaches to man and the environment. In *Fundamentals of human geography* (course D204), Unit 2, Section 1. Milton Keynes: Open University.

Rajasekaran, B, D. M. Warren, S. C. Babn 1991. Indigenous natural resource management systems for sustainable agriculture: a global perspective. *Journal of International Development* 3, 1–15.

Reed, D. 1990. *Structural adjustment and the environment*. London: Earthscan.

Richards, P. 1985. *Indigenous agricultural revolution: ecology & food production in West Africa*. London: Hutchinson.

Ruggie, J. G. 1975. International responses to technology: concepts and trends. *International Organisation* 29, 557–83.

Schroeder, R. A. 1993. Shady practice: gender and the political ecology of resource stabilisation in Gambian garden/orchards. *Economic Geography* 69, 349–65

Scoones, I. 1992. Land degradation and livestock production in Zimbabwe's Communal Areas. *Land Degradation & Rehabilitation* 3, 99–113.

Scoones, I. & J. Thompson 1993. *Beyond farmer first: rural peoples' knowledge, agricultural research and extension practice; towards a theoretical framework*. Mimeo, IIED, London.

Shiva, V. 1989. *Staying alive*. London: Zed.

Showers, K. B. & G. M. Malahleha 1992. *Historical environmental impact assessment: a tool for analysis of past interventions in landscapes*. Working Paper 8, The project on African Agriculture (Joint Committee on African Studies, Social Science Research Council/American Council of Learned Studies), Institute of Southern African Studies, National University of Lesotho.

Stocking, M. A. & L. Peake 1985. Erosion-induced loss in soil productivity: trends in research and international cooperation. *International Conference on Soil Conservation*, Maracay, Venezuela, 3–9 November.

— 1986. Crop yield losses from the erosion of alfisols. *Tropical Agriculture* 63, 41–5.

Taylor, P. & F. H. Buttel 1992. How do we know we have global environmental problems? Science and the globalisation of environmental discourse. *Geoforum* 23, 405–16.

Thompson, M. & M. Warburton 1985. Decision-making under contradictory certainties: how to save the Himalayas when you cant find out what's wrong with them. *Journal of Applied Systems Analysis* 12, 3–33.

— 1988. Uncertainty on a Himalayan scale. In *Deforestation and social dynamics in watersheds and mountain ecosystems*, J. Ives & D. Pitt (eds), 1–53. London: Routledge.

Tiffen, M & M. Mortimore 1994. *Land resource management in Machakos District, Kenya: 1930–1990*. World Bank Environment Paper 5, The World Bank, Washington DC.

Turner, M. 1992. Overstocking the range: a critical analysis of the environmental science of Sahelian pastoralism. *Economic Geography* 69, 402–21.

Vayda, A. P. 1983. Progressive contextualisation: methods and research in human ecology. *Human Ecology* 113, 265–81.

White, L. 1973. The historical roots of our ecological crisis. *Science* 155, 1203–207.

Wolf, E. 1972. Ownership and political ecology. *Anthropological Quarterly* 43, 201–205.

Zimmerer, K. 1991. Wetland production and smallholder persistence: agricultural change in a highland Peruvian region. *Association of American Geographers, Annals* 81, 443–63.

Zimmerer, K. 1993. Ecology. In *Conceptual thinking in human geography*, C. Earle & M. Kenzer (eds). London: Routledge.

CHAPTER TWO

Sustainable development: a challenge for agriculture

David Gibbon, Alex Lake,

Michael Stocking

Editors' introduction

"'Sustainable development' is an idea whose time has come." (Murdoch 1993). Not only was the concept of "sustainable development" one of the few tangible outcomes of the Rio *Earth Summit* in June 1992 (in the form of a Sustainable Development Commission), but it also now liberally peppers almost every policy document and project proposal in agriculture, environment, population growth and development. The term is applied to both developed and developing countries, to affluent and poor economies, and to temperate and tropical environments. It is presented by followers of ecodevelopment as the fundamental guiding principle (wced 1987). Strange to relate, then, there is still great uncertainty as to what it means, how it can be applied, and what practical outcomes it may have. As O'Riordan (1988) recognized, the very ambiguity of the term makes it attractive: developers like it because they can justify many an environmentally sensitive programme under its guise; environmentalists espouse it because it enables them to demand safeguards and compensating investments that may be economically unjustified and socially unjust. Nowhere does this lack of clarity and degree of confusion obscure the issues and challenges more than in the arena of agriculture.

Agriculture is expected to perform in two major ways, which many observers see as being mutually contradictory. On the one hand, agricultural systems have to develop and intensify to feed a growing population from finite land resources. They are expected to meet the aspirations and demands of the heterogeneous rural populations of the world now, while providing greater variety and quantity of foodstuffs to burgeoning urban populations. On the other hand, these same agricultural systems are the guardians of most

31

of the world's environments, expected not only to keep the stock of natural resources secure for current production but also to protect the "global commons" – the soil, air, water and biological diversity – for the needs of future generations. Can it be done? Is it feasible to expect agriculture to take on these multiple roles. Or are we asking the impossible?

In this chapter, David Gibbon, Alex Lake and Mike Stocking address the central issue of sustainability, pointing out its various facets and demonstrating that the term implies different outcomes from different perspectives. They contrast the problems of conventional agriculture – designed on a science-based paradigm of environmental manipulation, agribusiness and market forces – with the broader perspectives of environmental protection, stewardship of the land and low external-input models of agriculture. Inevitably perhaps, they conclude that there are no easy answers. Sustainability remains an overriding imperative, but there are many roads to it, each conditioned by the specific environments, resources and contexts of the persons attempting to ensure future food security.

Introduction

Populations make demands on the environment. They seek goods and services and "use" natural resources by tapping into global environmental systems in order to extract water, energy and nutrients. They expect the environment to support society, culture, wellbeing and, above all, development.

All such demands, whether social or economic, industrial or agricultural, necessarily have an impact on the natural resource base. These impacts are an inevitable outcome of the physical laws of thermodynamics: in the first law (conservation of energy), the sum total of energy[1] remains constant, so if something is extracted at one point, the stock of energy is depleted there but enhanced elsewhere; in the second law (the law of entropy), the status of the energy as it transfers becomes increasingly disaggregated and, hence, less useful to society, and it is only through the application of yet more energy flows (e.g. solar radiation) that concentrated energy and the stocks of renewable natural resources can be replenished.

---

1. "Energy", measured in basic units of joules and kilocalories, is taken in the wide sense to include solar radiation, primary energy such as the content of fossil fuels or growing biomass, delivered energy as in gas and electricity, kinetic energy as the energy by virtue of motion, and potential energy as the energy by virtue of the position of an object. To this could be added the energy of the atomic structure of particles.

Our effects on the environment may be likened to walking on shifting sand dunes – one moment the sand carries our weight and supports our steady progress, the next it collapses, causing the body to stumble. The natural systems have properties of "resilience" (an ability to absorb and utilize change, or withstand an external shock) and "sensitivity" (the degree to which change is induced, or how readily change occurs with only small differences in external force). Such notions of "resilience" and "sensitivity" give credence to the idea that some extractions and utilization of the natural environment are possible and reversible in their impact without undue loss; others are not, and the systems collapse (Table 2.1 provides two examples). If true, then our natural resources and environment can support some development but not all. The challenge is to know what development is permissible, and what development undermines our security and future.

Table 2.1 Resilience and sensitivity: a matrix for sustainability assessment of soil resources (after Stocking 1995).

Resilience	Sensitivity High	Low
High	Easy to degrade, but responds well to land management that restores capability	Only degrades under very poor management or persistent mismanagement
Low	Easy to degrade, unresponsive to land management and should be kept as near "natural" as possible	Initially resistant to degradation, but after severe misuse has great difficulty in restoring capability

Two examples:
1. The deep, yellow Loess Plateau soils of China have been cultivated for centuries. As evidenced by huge gullies and the sediment-laden Yellow River, they erode very easily. Yet, they are reasonably fertile and contain good reserves of nutrients. So local people carry on cultivating. *High sensitivity: high resilience.*
2. The oxisols of the Cerrado (savanna) in Brazil have only recently come under pressure for arable cultivation of wheat and soyabeans; before they were rangeland. These soils have an open structure and good physical properties. But intensive use of machinery eventually collapses structure and renders the soil very erodible. Restoration is very difficult. *Low sensitivity; low resilience.*

"Sustainable development" is the term coined in current debates to differentiate between development that is allowable and development in which the net impact is strongly negative and hence unsupportable. The term encompasses the flows of energy in natural systems upon which human society relies, and it should facilitate the assessment of resilience and sensitivity of the stocks of natural resources in order to achieve what at first sight may seem impossible: growth and development with security and protection.

Within the bounds of one chapter, a full review of the policy issues of sustainability and the economic difficulties of putting it into practice is impractical. Recent texts address this admirably; see, for example, Goodland et al. (1991), Meadows et al. (1992), Bartelmus (1994), Redclift & Sage (1994).

Sustainability: a question of debate

Throughout this century we have witnessed the continuing degradation of natural resources on a global scale, partly to exploit any resource that would lead to rapid growth and accumulation of wealth in the North. It has also been a consequence of our efforts to meet the food, fibre, fuel and other needs of the world's growing population, which is forecast to double in the next sixty years. Many authors (e.g. Blaikie & Brookfield 1987, Conway 1991, Meadows et al. 1992) argue that, as a result of an historical process determined by global political and economic decisions, the local management of natural resources has been not only conditioned by a variety of powerful external forces but also controlled by resource exploiters, end users and local residents. These influences have led inexorably to conflicts of purpose by the various "actors" who involve themselves with natural resources, ensuring in most cases that short-term exploitation succeeds over long-term protection.

Thus were early ideas of the need to conserve resources developed. Spurred by high-profile environmental disasters such as the American Dust Bowl of the 1930s, many parts of the colonially controlled developing world instituted conservation programmes. From 1953 to 1965, for example, the Federation of Rhodesia and Nyasaland had the largest single research programme in Africa on soil and water conservation. In international assistance for agriculture, the UN Food and Agriculture Organisation had by 1951, more than a hundred projects in 35 countries, and by 1959, 1700 experts in the field. Most projects were devoted to the assessment of soil, water and forest resources, locust control and other aspects of resource management aimed at increasing agricultural production through technical improvements (FAO 1992).

In 1960, the world's population reached 3000 million. For the first time, questions started to be raised about technology being able to address the exponential growth in human population, which at that time had a rate of increase of 1000 million people every 30 years. Thus, it was that the modern sustainable development debate can be traced back to the 1960s with the first United Nations Development Decade. Most observers then saw a solution to underdevelopment in the transfer of technology, expertise and finance to developing countries. The powering force was economic growth and the ideological stance had little room for the effects of such growth on future generations ("intergenerational equity"). The seminal contribution to the debate was the book *The limits to growth* (Meadows et al. 1972), which, in summary, said:

- If present growth trends in population, industrialization, pollution, food production and resource depletion continue unchanged, the limits to growth on this planet will be reached some time in the next hundred years. The most probable result will be a sudden and uncontrollable decline in both population and industrial capacity.
- It is possible to alter these growth trends and to establish a condition of ecological and economic stability that is sustainable far into the future. The state of global equilibrium could be designed so that the basic material needs of each person on Earth are satisfied, and each person has an equal opportunity to realize his or her individual human potential.
- If the world's people decide to strive for this second outcome rather than the first, the sooner they begin working to attain it, the greater will be their chances of success.

It was a persuasive analysis at the time, but it did attract criticism for being overly pessimistic. In their follow-up study 20 years later, Meadows et al. (1992) admit that some of their assumptions were flawed. They produced a parallel trio of statements:

- Human use of many essential resources and generation of many kinds of pollutants have already surpassed rates that are physically sustainable. Without significant reductions in material flows, there will be in the coming decades an uncontrollable decline in per capita food output, energy use and industrial production.
- This decline is not inevitable. To avoid it, two changes are necessary. The first is a comprehensive revision of policies and practices that perpetuate growth in material consumption and in population. The second is a rapid, drastic increase in the efficiency with which materials and energy are used.
- A sustainable society is still technically and economically possible. It could be much more desirable than a society that tries to solve its problems by constant expansion. The transition to a sustainable society requires a careful balance between long-term and short-term goals and an emphasis on sufficiency, equity and quality of life, rather than on quantity of output. It requires more than productivity and more than technology; it also requires maturity, compassion and wisdom.

These statements contain both the essence of the dilemma and the moral

35

imperatives for a sustainable society. They present the choices that need to be made within the not-too-distant future; choices that have never before been made on a world scale. Some might argue that humankind has never shown the necessary degree of social conscience to make the right choices, and we are therefore asking the impossible. Others would argue that, with the increasing regulation and control of manufacturing industry and of resource-use practices, we have already begun to respond to the Meadows' second and third goals. Yet others would argue that we are neither doing enough nor moving sufficiently quickly to avoid major environmental and resource depletion problems in the next century; the solution to the problems is possible but it needs to be urgently addressed.

An important change during this 20-year period has been an increased public environmental awareness. Many conferences, workshops, enquiry bodies and institutions have been held and set up to explore ways to slow down, halt or reverse the present rate of destruction and depletion of the world's resources. Key meetings and milestones in the sustainability debate may be identified between 1972 and 1992:

- The United Nations Conference on the Human Environment, Stockholm, 1972. One hundred and eight substantive recommendations were made to the world community. As noted by O'Riordan (1995), the Stockholm conference produced a conflictual relationship between North and South, eased only by the establishment of the United Nations Environment Programme with its headquarters firmly in the South (Nairobi). In the sphere of agriculture and rural development, habitat and resource conservation were addressed, as well as waste disposal and recycling technologies, food contamination control, and the monitoring of environmental problems associated with pesticide and fertilizer usage. UNEP was intended to monitor environmental changes and direct assistance to places where resource exploitation had created environmental damage. Sustainability was seen largely as a technological challenge.

- The World Conservation Strategy (WCS), 1980 (IUCN 1980). Reflecting its origin in conservation and environmental agencies, the WCS emphasized the interdependence of conservation and development and first gave common currency to the term "sustainable development". It specified three objectives related to genetic diversity, protection of ecological processes and life support systems, and the sustainable use of species and ecosystems.

- The World Commission on Environment and Development was formed in 1983 under the Chairmanship of Mrs Gro Harlem Brundtland, the Norwegian prime minister. It had a UN mandate to "identify and promote the cause of sustainable development" – a mysterious phrase that in effect allowed the Commission to range broadly in its published report, *Our common future* (Brundtland 1987). The Commission presented a powerful analysis of what was wrong with human occupancy of the Earth, as well as some strategies for fair access to basic resource needs for all people. One of its primary thrusts was the transfer of resources to the very poor through technology, capacity building and correct pricing of resources. Indeed, implicit in its view of achieving sustainable development is that the true costs to society of economic growth should be established, and resources should be safeguarded now as compensation for future generations.

- Caring for the Earth: a Strategy for Sustainable Living (IUCN/UNEP/ WWF 1991) was a follow-up to the World Conservation Strategy, compiled from the contributions of many international organizations and leading environmentalists through the IUCN Commissions. It presents "the reality that environmental, social and economic issues are joined in a network of sobering complexity" (p. 3) and stresses the "principles of a sustainable society". It is remarkable for its emphasis on the human condition and the implicit subservience of nature.

- The United Nations Conference on Environment and Development, the "Earth Summit", was held in Rio de Janeiro in July 1992. It was preceded by four two-week preparatory conferences involving 170 countries and 9 "stakeholder" groups representing indigenous peoples, environmental organizations, science, local government, business, farmers, trade unions, women and youth. The Conference itself attracted 110 world leaders, representatives of 153 countries, 2500 NGO groups and huge media attention. It was an amazing feat of organization and it raised the profile of the sustainability debate immeasurably. Politicians who never before had espoused conservation now waxed eloquently on environment and development. It gave the NGOs a greater confidence. Global conventions on biodiversity and climate change were signed. *Agenda 21* was, however, the main output of the meeting, setting out in 40 chapters sets of goals, action priorities, a follow-up programme and cost estimates. Nine chapters looked at the potential roles of the stakeholders. The major shortcoming of Rio was the failure to create international financing and organizational arrangements to deliver *Agenda 21*. The Global Environment Facility, created in 1991, has in part taken on the role of financial transferor with a first tranche to 1997 of US$1600 million. However, this is a far smaller sum than the developing nations demanded and UNCED calculated would be needed, and there is a mutual distrust by many parties as to the role of the World Bank as the main funding body. One of the more accessible guides to UNCED is by Grubb (1993).

Agriculture itself has tended to be somewhat sidelined in the sustainability debates of the past two decades. Concern over food security had declined somewhat with the huge world surpluses. The debate was largely driven by environmentalists and political activists who could see in the discussions the means of promoting particular causes, especially aid to the poorest and a Western-based view of science and the environment. However, not to be outdone completely, the international agricultural lobby through the UN Food and Agriculture Organisation has published two major reports:

- *Agriculture: toward 2000* (Alexandratos 1988) was FAO's attempt to look into the future development of agricultural systems, producing analyses of countries, commodities, technologies and resources within a context of increasing population and declining per capita resources. *Agriculture: toward 2010* (FAO 1993) updates that analysis with a focus on nutrition/food security and agricultural resources/sustainability. It describes as the very essence of sustainability the "safeguarding of the productive potential and broader environmental functions of agricultural resources for future generations, while satisfying food and other needs" (p. 1). Pessimistically, it predicts the continuation of chronic undernutrition, greater pressure on agricultural resources, and further challenges to sustainability.

What has been achieved by all these meetings, statements and voluminous reports? One view of the "UN conference-go-round" is that it is an "expensive circus of little relevance to the needs of those most desperate" (O'Riordan 1995: 21). Many of the pronouncements were indeed unrealistic, biased and somewhat naïve. The chapter in this book by Stocking, Perkin and Brown illustrates this very point in the case of biodiversity and attempts to link conservation with development. A more optimistic view would dwell on the increased awareness by all strata in society of the dangers of unrestrained growth, and a realization by governments, mainly in the North, that some environmental protection legislation is warranted, based upon proper "green accounting".

This chapter, therefore, looks at the challenges ahead in a climate where "sustainability" is very much the watchword. Inevitably, we have to grapple with the slippery definitions of sustainability and sustainable agriculture/development. Agricultural sustainability has generally been seen within the context of general concerns for the environment, the high use of non-renewable energy in production, the increasing dependence on external inputs and the requirement intensively to manage all natural resources. Surprisingly, perhaps, the criticism that followed the earlier euphoria about the benefits of the Green Revolution in the 1960s, was primarily concerned with social and political inequalities and very little on the longer-term potential damage to soil and other natural resources. One of the principal arguments presented here is that present concepts, strategies and methods of "modern" agricultural production would not appear to be sustainable in the long, and in some cases the medium, term (Conway 1985, Jodha 1990). Much of the thinking behind modern agriculture, which is driven by Western scientific ideas, policies of continued growth, artificially distorted markets and with a high dependence on non-renewable resources, is flawed as it does not take account of current or projected consumption patterns or long-term environmental and health hazards. At the same time, there is evidence from a series of cases that there are alternative strategies and routes towards more sustainable resource management and farming systems, which need serious study and consideration (NRC 1989, Reijntjes et al. 1992).

Our approach to the challenges ahead for agriculture will, of course, be predicated on our biases. Only by developing an understanding of individual situations, by allowing local knowledge to flourish, by understanding local cultures, values and institutions, and by combining these with scientific insights and more conventional ideas and practices, along with appropriate methods of experimentation and discovery – will sustainable agricultural systems be developed. It is a tall order. It will always be necessary to formulate policies and regulation mechanisms that support actions and knowledge systems that are compatible with more sustainable agricultural systems and take account of distributive and equity issues. However, the trend in our approach follows closely the writings of Brokensha et al. (1980) and Richards (1985).

Finally, to translate the approach into possible implementation strategies, we discuss some topical suggestions: the merits of low external-input agriculture, farmer participation in agricultural research and extension, producer–consumer groups, and communication networks. All of these have demonstrably shown that elements of sustainability can exist in today's agriculture. It is argued that the opportunities for a more independent and self-contained agriculture based on these strategies may be the means of achieving a more sustainable agriculture in both high and low productive potential environments.

Sustainability: a question of definition

Agricultural systems are complex and involve many people: farmers, other rural dwellers, consumers, conservationists, economists, politicians, women, youth and so on. The number of possible groups in society that interact with agriculture is immense. According to Brown et al. (1987), "Different societies have different conceptualizations of sustainability as well as different requirements for sustainability based on varying cultural expectations or environmental constraints". It is not surprising, therefore, that different perspectives prevail on the role of agriculture and the nature of sustainability. The different proponents have their own definitions, which reflect their perceptions, priorities, goals and value-judgements. Consequently, confusion reigns about what constitutes sustainable agriculture and what should be done about unsustainability. Many would argue that it is important to define what sustainability is, or might be, before any actions can be taken towards setting up more sustainable agricultural practices. We do not necessarily subscribe to the need to define sustainability in order to practise it, but the exercise of definitions is one useful way to examine several perspectives and to understand the competing views.

Sustainability takes on a role according to whether its definition is based upon social, economic or ecological perspectives, or a mixture of these. It is therefore important to specify the context, spatial and temporal scales that any definition of sustainability encompasses. Table 2.2 presents a sample of the many definitions of sustainability, sustainable development and sustainable agriculture, classified according to provenance. The diversity of emphases and degrees of precision is immediately obvious. It would be invidious to choose a "best" definition. Instead, below we summarize the major broad perspectives of sustainability, as viewed by the major proponents.

Global sustainability

The World Commission on Environment and Development, consisting of 22 eminent persons drawn from UN member States in both developed and developing countries, took a global perspective in calling for a "common endeavour and for new norms of behaviour at all levels and in the interests of all." Central themes of the Brundtland (1987) report, *Our common future*, were global co-operation and mutual support between countries at differing stages of economic development. The apparent simplicity of the Commission's definition of sustainability (see Table 2.2; and note that this is still the most widely cited definition) hides the complexity of achieving sustainable

Table 2.2 Definitions of sustainability, sustainable development and sustainable agriculture, categorized by perspective of the source.

Perspective, definition and source

Agriculture
- Sustainable development is "the management and conservation of the natural resource base, and the orientation of technological and institutional change, in such a manner as to ensure the attainment and continued satisfaction of human needs for present and future generations. Such sustainable development conserves land, water, plant and animal genetic resources, is environmentally non-degrading, technically appropriate, economically viable and socially acceptable." FAO (1991)
- "A sustainable land management system is one that does not degrade the soil or significantly contaminate the environment, while providing necessary support to human life." Greenland (1994: 3)
- "A cropping system is not sustainable unless the annual output shows a non-declining trend and is resistant, in terms of yield stability, to normal fluctuations of stress and disturbance." Quoted by Swift et al. (1991).

Economics
- Sustainability is "living on interest and not capital". Quoted by Bennett (1991).
- An optimal sustainable growth "policy would seek to maintain an acceptable rate of growth in per-capita real incomes without depleting the national capital asset stock or the natural environment asset stock." Turner (1988: 20)

Ecology
- Sustainability is "the net productivity of biomass (positive mass balance per unit area per unit time) maintained over decades to centuries." Conway (1987: 96)
- Unsustainability is "gambling with survival". IUCN/WWF/UNEP (1991: 4)
- Sustainable development is "improving the quality of human life while living within the capacity of supporting ecosystems." IUCN/WWF/UNEP (1991: 10)

Sociology
- "Sustainable (feasible) development can be advanced as the set of development programmes that meets the targets of human needs satisfaction without violating long-term natural resource capacities and standards of environmental quality and social equity." Bartelmus (1994: 73)

Composite
- Sustainable development is "development that meets the needs of the present without compromising the ability of future generations to meet their own needs" Brundtland (1987: 43)
- "Sustainable development refers to meeting human needs, or maintaining economic growth or conserving natural capital, or about all three." Redclift (1991: 37)

development in practice, a problem well acknowledged in the report. Consequently, the Commission tended to restrict itself to global sustainability and defined it as: "Meeting the basic needs of all and extending to all the opportunity to fulfil their aspirations for a better life", and it required that "those who are more affluent adopt lifestyles within the planet's ecological means". In their view then, sustainability is not a fixed and definable state; instead it is a process of change in which the exploitation of resources, the direction of investments, the orientation of technological development and institutional change are made consistent with future as well as present needs.

Sustainable development

Sustainability from the perspective of development is fraught with contradictions and has had to undergo some radical shifts in thinking (Elliott 1994). Of all the perspectives of sustainability, this is the one that is most invoked to support various political and social views.

The concept is wide open to such contrasting views. Take Conroy & Litvinoff's (1988) definition of sustainable development, for example:

> Improving people's material well being through utilizing the Earth's resources at a rate that can be sustained indefinitely . . . living off nature's interest rather than depleting capital.

Third World decision-makers can use this to copy Western development but at an accelerated pace on the grounds that self-evidently it was good for the West's material wellbeing, so why should it not be good for us? It is a difficult argument to refute without hypocrisy.

In the view of some commentators (e.g. Redclift & Sage 1994), sustainable development has gained common currency precisely because of the way it can be used to legitimize a host of different agendas. Indeed, the term is so useful that it can be used to construct local agendas of action as well as whole national policies – the case of the Philippines described by Remigio (1994) is instructive because, under the rhetoric of sustainable development, practices that in no way could be considered conservative (e.g. rampant deforestation, uncontrolled mineral extraction and dynamite fishing) are being promoted. Techniques in support of sustainable development, such as environmental impact assessment, provided no more than technical guidance on how to exploit, not whether to develop. Adams (1990) gives a good analysis of the "greening" of development, whereas Murdoch (1993) provides a useful theoretical discussion of the confusion about sustainable development, strongly advocating a new research agenda for the development of detailed strategies for the application of the concept.

Ecological/environmental sustainability

Original notions of sustainability have been largely driven by perspectives drawn from ecology and "environmentalism". From the biological sciences came the classic work on ecosystem functioning by H. T. Odum (1959), which publicized the interdependent complexities of natural systems, especially in regard to energy flows, and defined important new terms such as "ecological niche" (the status of an organism within its ecosystem). It was

42

but a small step to extend ecological principles to human interventions – and hence the development of human ecology - and whole habitats and agro-ecosystems. From these perspectives, then, sustainability becomes defined as the continuation of the functioning of natural biological processes.

In turn, these emphases on the interactions between organisms and their environment have led to some key concepts that have a high profile in the sustainability debate. Sustainable development is seen as encompassing the protection of genetic resources and the importance of biodiversity (see Ch. 6), presented often as a moral and ethical crusade for the rights of species to exist. Some writers (e.g. Capra 1980) believe that the increasing degradation of resources are symptomatic of crises in science and society. This view has undoubtedly spurred "deep ecology" – a broad philosophical spectrum of "green politics" requiring radical changes in social, legal, institutional and economic systems. Major changes in the quality and extent of human modification of nature are called for; there is an expectation of a return to pre-industrial, rural lifestyles – an impracticable proposition to most observers.

Closer to other perspectives of sustainability comes the concept of eco-development, which, in an interesting circularity of reasoning, is defined by Brown et al. (1987) as "ecologically sound development, a process of positive management of the environment for human benefit." According to Colby (1989), eco-development explicitly sets out to restructure the relationship between society and nature into a "positive-sum game", where both sides gain by appropriate development, through sophisticated forms of symbiosis. Most sustainable development activity would be the management of this relationship between society and ecological processes, incorporating social equity and cultural concerns, while preventing resource depletion and promoting "green growth". The potential contradictions in such perspectives are many (see Redclift 1987, Adams 1990).

Economic sustainability

The concept of economic sustainability differs from the other perspectives in that lower priority is given to ecosystem functions and resource depletion. Positive economic growth is seen as possible within the bounds of a fixed total capital stock. A useful three-level perspective on sustainability is given by economists Pearce & Warford (1993) and adapted by O'Riordan (1995):
- Weak sustainability: overall stock of capital assets (natural, human and man-made) remains constant; sustainability seeks the substitution of one form of capital for another

- Moderate sustainability: constant stock cannot be maintained because of constraints imposed by assimilative capacity; the critical natural capital has to be protected from irreversible decline or catastrophe
- Strong sustainability: guards the primacy of ecosystem functioning in the most cost-effective and "natural" way; nature is protected for a balance of utilitarian, precautionary and ethical reasons. Indicators used: existence value, bequest value, aversion to uncertainty, tolerance to depletion, environmental wellbeing. A premium is placed on peace of mind in knowing one lives in a sustainable world!

These economic perspectives of sustainability have gained a primacy by virtue of the institutions that promote them and the importance ascribed to economic considerations in investment decisions. For developing countries, the approach adopted by the World Bank is crucial. Recent trends in actions towards sustainability have included, for example, "polluter-pays" principles, where those who cause the pollution bear the costs of rectifying the damage. On a larger scale, international environmental accords now espouse the cause of compensatory balancing payments for greenhouse gas emissions and "carbon taxes". The European Union has suggested that such taxes would finance research and development into further technologies to reduce emissions. Another example of the economic perspective in action is the proposal of debt-for-nature swaps, or the trading of developing countries' huge debts on a secondary market to pay for conservation. The first such swap occurred in 1987 when Conservation International paid US\$100000 for US\$650000 worth of Bolivian debt, in return for US\$250000 in local currency towards the Beri Biosphere Reserve. Such swaps are now also on the agendas of the major development banks.

Monetary valuation of environmental goods and services towards a goal of sustainability thus assists the making of choices. The wisdom and rationality of these choices depend on the way the valuations are carried out, and there is much cause for scepticism over the underlying assumptions used by environmental economists and the actual values so derived (Bartelmus 1994). We are left, however, with the problem of how to price the priceless. It is clear with techniques such as contingent valuation and integrated accounting that an economic perspective of sustainability is bound to stress particular aspects of development: for example, infinite economic growth and prosperity; exploitation of resources controlled by the market; property and resource ownership regimes along with privatization; an emphasis on anthropocentric problems such as hunger, disease and poverty; and environmental management technologies such as pollution dispersal, biotechnology and high inputs of energy.

A comparative view of sustainability

It is tempting to conclude this section of definitions of sustainability with the observation that the word means all things to all people. There are, however, some common principles that should be included in any operational view of sustainability:

- continued support of human life
- the right of future generations to access to resources of equal worth
- long-term maintenance of the diverse stock of biological resources and products of agricultural systems
- stable human populations in economies with limited or controlled growth
- an emphasis on greater autonomy and self reliance
- continued maintenance of environmental and ecosystem quality.

The various facets and perspectives of sustainability do, however, bring out a richness of possible means of achieving the goal of sustainable development. Table 2.3 illustrates a view of the extremes in the debate, comparing an economic worldview perspective with "deep ecology". With such a divergence of fundamental objectives in support of actions towards sustainability, there is little wonder that confusion reigns. This chapter now examines how agriculture has responded to the complex roles demanded of it by the challenge of sustainability.

Table 2.3 Extremes in the range of views of the guiding imperatives in sustainability: the dominant economic world view versus "deep ecology" (adapted after Colby 1989).

Dominant economic worldview Major proponents: development banks, national aid agencies	Deep ecology worldview Major proponent: *The Ecologist* magazine
1. Dominance over nature	1 Natural harmony; symbiosis
2. Natural environment is a resource for humans	2. All nature has intrinsic worth; biospecies equality
3. Material and economic growth for increasing human populations	3. Simple material needs; goal of self-realisation
4. Belief in ample resource reserves	4. Earth's resources limited
5. High technological progress and solutions	5. Appropriate technology; non-dominating science
6. Consumerism & growth in consumption	6. Do with enough; recycling
7. National/centralised community	7. Minority traditions & indigenous community knowledge

Agricultural sustainability

As defined by one of the new arch-pragmatists, Joaquim Chissano (President of Mozambique), to the UNCED Conference at Rio, a sustainable agricultural production system is "one that indefinitely meets rising demand for food, fibre, and other agricultural commodities at economic, environmental, and other social costs consistent with rising per capita welfare of the people served by the system." (Chissano 1994: 37) The economic perspective has tended to imbue agriculture in its quest for sustainability. Chissano's definition picks up two key points that have exerted powerful influences in agriculture: first, the recognition of the production–growth linkage; secondly, the necessity for improvement in livelihoods as a consequence of human endeavour and by means of agricultural production systems.

A useful way of articulating the basic components of agricultural sustainability is given by Douglass (1984):

- through sustainability as long-term food sufficiency, which requires agricultural systems that are more ecologically based and that do not destroy their natural resources
- sustainability as stewardship; that is, agricultural systems that are based on a conscious ethic regarding humankind's relationship to future generations and to other species
- sustainability as community; that is, agricultural systems that are equitable.

The problem with all such component lists is that the means of achievement are not specified. Consequently, the practical reality of implementing sustainability can be rationalized in almost any agricultural system. Conventional agriculture through capital intensive investment epitomizes a technological approach to achieving rising per capita welfare. Only the most hard-bitten of intensive commercial farmers would now accept that conventional agriculture is sustainable; nevertheless, it trades strongly on 1960s versions of sustainability through two primary strands: the nature of Western science, and the global agribusiness economy. Each of these threads will be examined, because they have been (and continue to be) extremely powerful forces in decision-making in agricultural practice.

The nature of Western science

"Scientism" describes an approach to knowledge that pre-eminently promotes the results of science as a basis for decisions and irrefutable fact.

Western society, along with agricultural science, has become dominated by scientism as the only accepted mode of analysis and understanding. It has affected intimately the way that science has served development in developing countries (Abel & Stocking 1981). This type of knowledge places emphasis on reductionist ways of regarding nature and the physical world, and has led to the isolation of academic disciplines, and the traditional single disciplinary government and agency structures for professional advice. According to Capra (1980), "it has served as a rationale for treating the natural environment as if it consisted of separate parts, to be studied and exploited by different interest groups". This value mindset of emphasis on scientific method, and Cartesian rational analytical thinking, has resulted in "profoundly anti-ecological attitudes". This is because much of scientific thinking involves a linear process; the understanding of ecosystems with their "non-linear dynamic balance based on cycles and fluctuations" requires a certain amount of intuition and understanding of nonlinear systems. Capra argues that Western society has, to a certain extent, lost this intuitive ability of enquiry, but that it is still a very prominent and legitimate mindset in many Eastern and older societies in developing countries.

These persuasive arguments about the nature of scientific thinking continue to be subject to lively debate. There is no doubt that agricultural scientific training remains dominated by a few influential Western scientific educational institutions and has a character that has changed little in the past 40 years. We continue to train natural scientists without a good understanding of the social and political context in which they work and the role they themselves play in determining outcomes. It is a painful process for both developed and developing countries to accept the post-Rio *Agenda 21* recommendations: of the 115 programme areas for action set out in the document, some 70 are relevant to global agriculture and the work of agricultural researchers in developing countries. It is by no means certain that governmental and academic institutions can adapt and be able to integrate sustainable development themes into their agriculture programmes. A recent (1993) FAO survey found a wide divergence in practice.

The agribusiness economy, demand and consumption

Capital intensive agriculture has been developed in the West and in some parts of the tropics to serve the needs of growing markets and the demand for an adequate return on investment in capital, land and labour. Such agriculture has evolved in direct response to the prevailing economic climate and to policies that have generally pushed to maximize production through national and international subsidies or in response to multinational company agendas. High-input farming practices inevitably predominate, and agricultural sustainability becomes a matter primarily of continuing to satisfy a production quota or a demand: real, imaginary or created. The focus is on the current economic situation rather than the physical availability of resources:

> Physical limits of resources almost never bar increases in the output of particular products or sectors of an economy as long as society is willing to incur the economic and/or environmental costs of recruiting the resources from other potential uses.   (Douglass 1984)

There is a lack of concern about environmental costs arising with the expansion of food, or agro-industrial supplies. The negative effects/externalities of farming are discounted into the future; therefore, they do not see it as being desirable to minimize environmental costs at the expense of current economic costs. It is assumed that technological advances can make up for effects of resource depletion or damage with new scientific discoveries providing solutions to new problems as they emerge. Hence, high expectations of scientific agriculture are built up, and extravagant claims are made for some of the more sophisticated branches such as biotechnology.

Such science-based systems are grounded on the same optimism that drove the Green Revolution in the 1960s in developing countries and which resulted in spectacular gains in production of cereal grains in the high-potential areas of several countries that were hitherto net grain importers. There is now considerable concern about the levelling off of production in these systems and the serious health, water-quality and soil degradation problems that are emerging (Harrington & Hobbs 1992). Evidence from these studies would suggest that the era of significant annual gains in crop yields in high potential areas is over and that a range of alternative strategies need to be examined.

Alternative agricultural systems and approaches

It has been widely accepted that, "to have an ecologically and economically healthy world, we have to make changes in agriculture" (Firebaugh 1990: 674). Such feelings, driven by a strongly held ecological and environmental philosophy, have driven the alternative agriculture movement into radical departures from science-based and technological agriculture. Derided by some as the "muck and magic" school, alternative agricultural systems have been a deliberate attempt to return to production based upon the functioning of natural systems and the utilization of ecological processes, rather than gross manipulation of nature and high external inputs of energy. These simpler agricultural systems do, indeed, have a sound basis in energy efficiency. Table 2.4 illustrates, for example, the 50-times greater efficiency of production output of a peasant farming, largely organic, system than intensive spinach cultivation in the USA. If transport and processing energy are also included, energy efficiency declines further in developed economies. An extreme example quoted by Sousan (1992) is the can of "diet" soft drink with its one kilocalorie of energy, but which takes 2200 kcal of energy to produce. In such cases, the energy content of the product is completely trivialized.

Table 2.4 Energy efficiency of agricultural production systems (kcal per ha per year; based on data in Pimentel 1984).

Agricultural system	Total energy input (A)	Energy production (B)	Energy efficiency (B/A)
Hunter/gatherer	2685	10500	3.9
Pastoralism (Africa)	5150	49500	9.6
Peasant farming(Mexico)	675700	6843000	10.1
Estate crop production (Mexico)	979400	3331230	3.4
Estate crop production (India)	2837760	2709300	0.9
Maize (USA)	1173204	3306744	2.8
Wheat (USA)	4796481	8428200	1.7
Rice (USA)	14586315	21039480	1.4
Apples (USA)	18000000	9600000	0.5
Spinach (USA)	12800000	2900000	0.2
Tomatoes (USA)	16000000	9900000	0.6

Following from the implications of the data in Table 2.4, sustainable agriculture becomes a loose set of strategies that look to utilizing energy efficiently. Much that has been promoted as alternative agriculture embodies these principles of energy utilization, but it also includes some key physical and biological criteria. Based upon the results of pragmatic experiences in US alternative agriculture over the past 15 years, a useful list of these criteria is provided by Lockeritz (1988):

- diversity of crop species
- selection of crops and livestock that are well adapted to particular environments
- preference for farm generated resources rather than purchased materials
- use of nutrient cycles to minimize nutrient losses
- livestock housed and grazed at low stocking densities
- enhancement of storage of nutrients in the soil
- maintenance of protective cover on the soil

- rotations that include deep rooted crops and which help control weeds
- use of soluble inorganic fertilizers
- use of pesticides for crop protection only as a last resort.

With the exception of the last two practices, most proponents of alternative and low-input agriculture would agree with this set and would include the practices as part of ecologically based agriculture (e.g. Rodale 1983, Altieri 1987, Haverkort et al. 1992). Others adopt more radical variants that strive for "pure" systems that cannot integrate effectively (and often deliberately) with conventional farming. For example, the long-established biodynamic movement (Steiner 1924), natural farming (Fukuoka 1988) and permaculture (Mollison 1988) are attempts at adopting alternative philosophies of interaction with nature, which set aside the dominant production goals of modern agriculture. Most of these idealized systems do not contain social, economic or policy dimensions, but trade rather on strong ethical and moral arguments. They are, in effect, a long way towards the "deep ecology" perspective of sustainability, discussed earlier, as applied to agricultural practices (see Dahlberg 1984, Wiseman et al. 1991).

Anathema to alternative agriculturalists is the intensive use of artificial fertilizers, herbicides, fungicides and pesticides. However, it remains clear that wholesale change in land-use practices will not occur until there is clear evidence that shows that current practices are unsustainable or that they present a serious risk to health. Many biological research scientists, advisers and farmers would similarly support this view. That there is already much evidence available to substantiate the damaging effect of high-input practices (Clunies Ross 1992) is largely ignored. Those that do heed the warnings, such as the growing number of people in the organic and biological farming movements and those who have an holistic view of nature, would consider that the moral and ethical factors outweigh the "scientific" and current economic evidence. These are the very people who are convinced that there already are serious problems and that world agricultural systems need to switch immediately to less environmentally damaging practices.

Broader perspectives

The narrow biological/ecological concept of sustainability is limited to the farm systems level, because it is the level that can be most easily tested. However, with the world becoming increasingly more interdependent, "sustainability" takes on different components, requirements and meanings, according to whether global, national and local levels are being considered. In addition, all these levels have different sociopolitical, ethical, philosophical and legal implications that cannot be quantified in a scientific manner but may be the most important to consider in the move towards more sustainable agricultural systems. These dimensions are seldom a priority in the mandates of either national or international agricultural research centres. Agricultural researchers operating in such centres often disregard externalities of a sociological or cultural nature because they involve value-judgements. They insist on determining sustainability primarily at the level of biological and physical efficiency.

Okigbo (1989) considers that this is unacceptable and declares that "sustainability can only be achieved when resources, inputs, and technologies are within the capabilities of the farmer to own, hire, maintain and manage with increasing efficiency to achieve desirable levels of productivity in perpetuity with minimal or no adverse effects on the resource base, human life and environmental quality." This broader and more pragmatic perspective is clearly important for agriculture in developing countries. An interesting example of its application is given by Giller et al. (1994), when they ask, "Can biological nitrogen fixation sustain tropical agriculture?" They answer "yes", but put conditions on the delivery of a wholly biological N_2-fixation system: adjustment to lower outputs, inclusion and acceptance of legumes in pastures, involvement of farmers in the feasibility studies and actual design of recommendations, and a more thorough devotion to suitable extension methods. In other words, it is non-technical factors that are crucial to generating sustainable tropical agriculture. The following sections look at some features of sustainability that may help to fix some important attributes of the concept in today's agriculture: unsustainability and the use of indicators, recognizing the role of local and indigenous knowledge, stewardship and community concepts, and developments in farming systems research.

Unsustainable systems and sustainability indicators

Jodha (1990) approaches the issue of agricultural sustainability from the perspective of farming systems in hazardous mountain and marginal environments. He suggests that, since agricultural sustainability is so complex and difficult to define, it is more logical to look at the concept of unsustainability. He claims that it is possible to examine, observe and measure the factors and processes that contribute to unsustainability, then prepare an inventory of indicators of unsustainability in any system and the reasons for their presence. Jodha has concentrated on what he calls the "fragile resource zones" that are the most vulnerable to inappropriate, unsustainable farming

51

practices. He argues that in the past, before populations began to increase significantly and external influences began to intervene in local farming systems, older, indigenous, farming systems were sustainable for centuries.

There has been a great deal of interest generated recently in the idea of measuring system characteristics and developing sustainability indicators (Harrington 1991, Verbruggen & Kuik 1991) and in devising methods of combining an historical analysis with social, economic, physical and biological indicators (ten Brink 1991). Sustainability indicators range from field-level assessments of soil quality, land degradation and vegetation variables, to macro-level parameters of food security and economic efficiency. If an agreed and appropriate set of such indicators can be developed from a range of case study analyses, it may be possible to develop operational tools that can be widely applicable. No such set yet exists, but scientific and policy advisors at the World Bank, UN Development Programme and bilateral development agencies are pressing ahead in an attempt to test many indicators for their utility in decision-making. This work is likely to see some significant advances in the next few years.

Local and indigenous knowledge

Indigenous farming practices and systems of agriculture have evolved through generations of informal experimentation and are now recognized as worth serious study. These practices are often better adapted to the limitations and potentialities of fragile environments and are therefore important to understand (Altieri 1987, Chambers et al. 1989, Jodha 1990). A good example from India in local soil and water conservation techniques comes in an advisory booklet *Farmers are engineers* (Premkumar 1994) that sets out in no uncertain terms how farmers' own experimentation may come to surprisingly different recommendations to that of agricultural extension; for example, the use of earth bunds as silt traps to extend cultivable area.

Frequently this kind of local agricultural innovation may exhibit some or all of the following characteristics:

- small scale
- diversified and interlinked structurally
- extensive or intensive land management
- use of locally renewable resources
- supported by folk knowledge and oral tradition
- control through social sanctions and local policing
- low use of external inputs
- low levels of total production and less genetic specialization

- autonomy and decision-making at farm level
- a strong focus on sustainability and food security.

These characteristics are generally conducive to sustainable resource use under low populations, but may not be viable under increasing pressure on land resources. However, there are examples of systems that have shown considerable resilience and innovation, without the addition of large quantities of external inputs, even in the face of rising population densities. The example of locally developed changes of agricultural land use in Machakos (Kenya) is instructive in showing how rising populations may provide the spur to innovation – a case of change and survive, or stagnate and die (Tiffen et al. 1994). Increasingly, these types of modifications to farming are being recognized from both the tropical developing world and in developed economies, such as those that occur in Europe (van der Ploeg 1990, Reijntjes et al. 1992). There are, of course, also many places where innovations do not seem to occur and where local systems break down in the face of major natural disasters or wars, migrations and other human-induced stresses. For example, The Horn of Africa has seen major conflicts and the abandonment of agricultural systems. However, this is not to say that over time new systems will not develop in response to new conditions; indeed, there is evidence that such systems are even now arising in Ethiopia as stability starts to return to places such as Eritrea.

Sustainability as stewardship

Farmers use up both renewable and non-renewable resources. In the long run, nature imposes limits on the amount of resources used to provide food for the world. Subsidies from renewable resources will soon dry up, so sustainability, in this sense, depends on the availability of a renewable resource base, and control on demands of its output that will insure against its depletion. This is stewardship, or working within the sustainable yield capacity of renewable agricultural resources. A useful definition of sustainable agriculture in this respect is provided by Brown (1984): "An habitual tendency to use natural resources in a manner respectful of their own beauty and harmony, that also conserves their aesthetic and productive values". Stewardship is seen as the key to a rational balance between resource use and food production; it is the manipulation of resources only to the degree necessary to satisfy human needs. He argues that research supporting "prudent public conservation" (or stewardship) will rarely command public interest and funding support because of lack of immediate economic interest.

Stewardship devolves responsibility to the farmer. Herein lies a difficulty in that land users may be unwilling or unable to carry out what are understood to be more sustainable actions attributable to lack of resources, inadequate economic returns or insufficient information about the longer-term or wider benefits. In both developing and developed countries, stewardship is frequently overshadowed by the priority of increased production: "Sustainable agriculture in the developing world cannot be attained simply by supporting sound land and water conservation techniques" (Brown 1984). It therefore has its practical limitations.

Sustainability as community

The most distinguishing feature in relation to sustainability as community is the focus on values of the community and the sense of belonging to a group (Pretty & Sandbrook 1991). This approach rejects the idea that the primary purpose of agriculture is production and growth. "Efficiency" in farming is redefined as looking far into the future and adopting methods of production that preserve the source of nature's bounty (Douglass 1984).

"Modern agriculture" has difficulty in addressing this view of sustainability. Traditionally, the identity of the community was located in the geographic proximity of individuals, reinforced with ties of kinship and patterns of common agricultural experience. Now, community becomes a sense of shared objectives in maximizing returns; but this means exploitation of neighbour and hence is contradictory to a sense of community. With the increasing specialization of modern agriculture, new kinds of communities have formed on the basis of class, wealth and common economic interests. On account of new measures of status and success (wealth, prestige and power), the diversity of experience and mutual interdependence in older communities has often been lost and replaced by new kinds of linkages and networks and a greater degree of dependence on external support. Nevertheless, these dependencies may forge a sense of community or solidarity that in turn supports agricultural sustainability.

Farming systems research and beyond

Important developments in farming systems research and extension (FSR/E) have aided the development of more effective interaction with farmers and with communities (Shaner et al. 1982). This has included the acknowledgement of the importance of understanding farmer knowledge (Brokensha et al. 1980, Richards 1985) and of major changes in the roles, attitudes and approaches to on-farm research. Shifts in attitudes and thinking have been stimulated by the emphasis on farmer-first and farmer-participatory research (Farrington & Martin 1987, Chambers et al. 1989). This work, in turn, has contributed to an urgency to develop methods of rapid rural appraisal (RRA) in support of finding local complexities without elaborate, costly and long surveys (e.g. Conway 1985). It was but a small step to bring RRA together with the evolving consensus on including farmers themselves in their own analysis into participatory rural appraisal (PRA). Methodologies of PRA and several variants (see Cornwall et al. 1992) are now stimulating considerable interest among many in rural development research and agricultural extension. The danger is that they are also being seen by some as a quick solution to complex problems, leading possibly to inappropriate or unsustainable solutions.

In a recent review of the theoretical and applied literature on farmer participatory research, Okali et al. (1994) suggests that the major challenge in relation to the explosion of interest in PRA activities is the need for researchers to develop a clear understanding of the nature of local knowledge and the farmer-led experimentation process. Without this understanding, no matter how many rapid techniques, methods and tools are deployed, it is doubtful whether any sustainable benefits can be achieved. Further, they highlight that rapid techniques may be being used as a substitute for careful, long-term observation, measurement and monitoring of rural-change processes with the consequent loss of vital information obtainable in no other way.

Values, perceptions and priorities of sustainable agriculture

"Values are the key to the evolution of a sustainable society, not only because will they influence behaviour but also because they determine a society's priorities and its ability to survive" (Brown et al. 1987). If values are the guiding force of the actions of society, any activity that aims for a more sustainable agriculture must take into account the nature of that particular society for which the activity is designed. It cannot be assumed that the values and the value-judgements of external professionals will necessarily and automatically prescribe actions that are compatible with the values of local groups. Indeed, it has been quite the reverse experience in many aid programmes in developing countries, which have failed precisely because the values of the designers were different to those of the aided.

Values are also the starting point for a change in our own agricultural practices, which, with the production of surpluses and increased pollution as a result of the use of modern farming methods, need radical change, just as much as many developing country agricultural practices need changing. In the view of a growing number of observers, Western society has become separated from nature, leading to more abuse of natural resource endowments (e.g. Capra 1983).

However, there are signs that the singular and unrelenting pursuit of materialism is beginning to wane. Many of the more affluent in Western society are beginning to practise what Brown (1984) calls "voluntary simplicity", a lifestyle that aims at the acquisition of goods only to satisfy basic personal needs and which seeks satisfaction from personal development rather than material acquisition. Herman Daly (as cited in Brown 1984) has observed that turning our focus to meeting basic needs will make fewer demands on our environmental resources but much greater demands on our "moral" resources. It may be reasonably assumed that value changes will, within time, translate to shifts in public policy and a re-ordering of priorities. However, it may be a long time before the imperatives of economic growth and advancement change sufficiently to allow a "values-driven" society to determine priorities that include sustainability in the use of resources. Although Western societies may be starting to turn towards new values, emerging economies are even now discovering the "virtues" of rampant capitalism; it may take many generations, several eco-disasters, and a reawakening of spiritual and moral values, before there is worldwide consensus on sustainability.

Truly appropriate ecological awareness will arise only when intuition is combined with rational knowledge. Since the cultivation of intuitive wisdom in Western culture has been neglected, there is a compelling case for respecting indigenous and locally developed knowledge along with their value systems. How can scientists and researchers possibly formulate solutions for agricultural sustainability, which fundamentally must internalize a wealth of location-specific characteristics, if they have no intuitive connection to the area they are dealing with? Problems concerning nature and human interactions cannot be solved by linear thinking alone. Systems thinking is one important way that has been developed to integrate the complexities, and low external-input agriculture a way to realize the intrinsic qualities of natural systems; these are addressed in the following two sections.

Systems thinking

As an alternative to reductionist science, systems thinking has produced many theoretical contributions to complement more traditional approaches to scientific enquiry. This thinking views the world in terms analogous to ecosystems; that is, the pre-eminence of the interrelations and connections between elements within systems. A system is an integrated whole, the properties and outputs of which are different from those of its constituent parts. Systems are also considered as having boundaries defined by the user, which may be distinct or deliberately fuzzy; they have inputs and outputs that may be measurable; they have internal workings that may be observable and measurable ("white" and "grey" box systems) or completely unknown ("black" box). Hard and soft systems have also been defined (see Checkland 1988), each of which assists the analysis and understanding of component interactions of different types of system problems.

As with the previous discussion on the importance of alternative values to material accumulation, systems thinking has a vital potential role in sustainability through the promotion of different ways of analyzing the real world. Only holistic modes of analysis have the capability and rationality to incorporate material dimensions with the philosophical, spiritual and moral arguments that have to underpin the concerns for the future, which are so much a part of the sustainability debate.

Low external-input agriculture

One way of realizing in practice a flight from a materialist-based culture is to return to a reliance on the properties and functioning of natural systems. Agriculture then becomes the means whereby the natural systems are tapped for a moderated amount of energy, goods and services. These ideas are incorporated into low external-input sustainable agriculture (LEISA), a movement promoted by the Information Centre for Low External-Input Agriculture in the Netherlands (ILEIA). Parallel movements exist in other countries: for example, LISA in the USA. ILEIA has become influential in some developing countries, primarily through non-governmental organizations. It has fostered a focus on the increased reliance on local self-sufficiency, in terms of not only physical inputs but also peoples' labour activity and indigenous knowledge (ILEIA 1988). LEISA means less reliance on external technical and material inputs, markets and knowledge, all of which have frequently been proven to be problematic because of a dependency on risky and uncertain supplies (Reijntjes et al. 1992). The marketplace is still important, but an overreliance on distant markets over which the producer has no control is discouraged in the LEISA approach. Is this return to a more basic agriculture, reliant on natural endowments, a feasible proposition for agriculture, which is expected to be able to feed growing populations as well as to be sustainable? The answer has to be a cautious "yes". Developing countries can no longer afford to suffer the damaging interventions of technology developed for non-local situations, research undertaken with foreign values and perspectives, and assistance with a top-down approach. IUCN/UNEP/WWF (1991: 113) give a useful listing of components of good land husbandry, none of which relies on externally provided material sources:

- *Respect land capability.* Use of the land must be matched with its suited purpose. Indicators of suitability and capability will be evident in how stable local societies currently use the land.
- *Conserve soil.* Sustainable practices generally improve the organic-matter status of soil and prevent the unnecessary depletion by erosion of stocks of nutrients. They also use the inherent capacity for recovery of natural systems through the biological activity of the upper soil layers.
- *Manage rainwater.* For most marginal areas in the tropics (even in the humid tropics), the conservation of water is a key to sustainable agriculture and production. In other situations, waterlogging and associated degradation processes such as salinization and alkalinization are major challenges to agricultural production. Plant yields are reduced more by water stresses or by excess water than they are generally by nutrient deficiencies.
- *Maintain plant cover.* The protective role of live and dead vegetation is well known: a mulch cover is the most effective means of minimizing soil and water losses. Ancillary roles include nutrient storage and promotion of more equable microclimates. The more prone to erosion is an area, the more important is the maintenance of dense and long-lasting cover: trees, perennial crops, plant mixtures (intercropping and agroforestry) and ground-mulching systems such as conservation tillage and green manuring.

In addition to these priority criteria for land-use practice, IUCN/UNEP/WWF (1991) stress:

- the promotion of co-operation between technical staff and local communities, and
- the adoption of practices that are both productive and conservative.

In this regard, Lundgren et al. (1993) provide a useful set of experiences from Africa of projects supported by SIDA (Swedish Aid), which have varied considerably from place to place, but all of which are based on capacity-building of local communities, self-reliance, institutional change and mobilizing local people as the primary agents for change. They give, for example, the case of the Dosso Region of Niger, characterized by low and erratic rainfall, poor soils and low water table. The low external-input approach to land use and the involvement of the external aid agency is centred around three key factors. First, a scheme to provide credit to individuals for activities related to production is managed by elected committees within villages. Secondly, the villagers undertake an annual self-evaluation of the project, with strong encouragement to be frank and business-like and to deal swiftly with grievances and injustice. Thirdly, joint training of villagers and officials is undertaken, dealing with technical, process and management issues.

Of course, none of these measures by themselves is a panacea for sustainability. However, they accord closely to the findings of a recent review of research and development in sub-Saharan Africa (Haverkort et al. 1991) and a review of alternative agriculture in the USA (NRC 1989), that many agencies, institutions and individuals are concerned with environment and development issues and are attempting to develop agriculture and natural resource management techniques within a long-term, more sustainable perspective. These reviews also show that alternative systems cannot work effectively without an appropriate supporting policy framework that includes appropriate training and communication.

Training, communication and networks

Although major contributions have been made in the development of farming systems research and extension (FSR/E), there has been relatively little impact so far within agricultural research or agricultural education institutions (Stocking 1994). FSR/E has itself evolved rapidly from contributions based on conventional hard systems thinking into more flexible approaches based on participation (see Baker & Norman 1990). Developments in this area have suggested that there is a need to rethink the structure, orientation and style of agricultural educational institutions. The Faculty of Agriculture of the University of Western Sydney (Australia) is one example of an institution that has taken on board these principles and has translated them into new structures and experiential learning programmes within agricultural education (Bawden 1990). We would contend that, without this fundamental rethink of the structure, organization and orientation of basic educational institutions, little progress will be made in generating the diversity of options required to deliver sustainable agriculture.

Hanson & Regallet (1992) see sustainable development primarily as a challenge for training and communication methods. The move away from human intuition and community awareness that modern Western society has been experiencing for the past century is for such authors the main block to the process of sustainable practices. They ask why the power of modern communication and information systems seems to have been so little used to stimulate the thinking and action required for the Earth to remain a sustainable home for the world's peoples. This concern is echoed by Richard Bawden's group in Australia (see previous paragraph) who articulate the need for "sustainable conversations" between and within all groups in society, to include rural people, town-dwellers, young and old, male and female, lecturer and student, theoretician and field worker, industrialist and agriculturalist, and any other people who utilize Earth's resources. Consequently, there have been recent trends towards improving communication between different groups who have a concern for sustainable agriculture, and many networks have been set up to serve the needs of alternative and sustainable agriculture: ILEIA, IFOAM, GRET, AGRECOL, AFSRE, ODI, AGRINET and many local NGO forums. The concept of "networking" has arisen to facilitate the more efficient exchange of information about methods, theory and practice of sustainable agriculture. Electronic mail is now being used, as in, for example, the SARD (Sustainable Agriculture and Rural Development) network and bulletin board, co-ordinated from the UNDP. INFORUM is another electronic network for the support of sustainable agriculture: in February 1994 it co-ordinated a three-month international "electronic conference" on the topic of sustainable indicators with 500 network participants. An important aspect to note about networks is that their members participate on a voluntary basis, they carry out joint activities that cannot easily be done alone, their individual autonomy remains intact, and the network's structure is usually informal and encourages participation.

The potential value of local networking in the search for agricultural sustainability is considerable. It can provide a very necessary step towards greater interdisciplinary co-ordination amongst professionals, as well as vertical communication from the farmer/land-user to the policy-maker. It draws together a wide range of individuals and organizations who are scattered but are working towards common goals.

Conclusions

There are no universal solutions or simple formulae for agricultural sustainability, nor should we expect to find any. Agricultural sustainability is a subset of the whole sustainability debate, and it includes the diversity of perspectives and often conflicting goals of different groups in society. The concept itself will undergo constant change in response to changing economic climates and environmental concerns, but there are some common principles and identifiable needs that sustainability embraces.

First, there is the need to develop an agreed set of indicators that incorporate environmental, economic and social concerns and which could be tested in a range of different circumstances. The outcomes of this might be, as a minimum, a measure of agreement on the limits of resource exploitation over time, and important interdisciplinary perspectives on sustainability and unsustainability. At an academic level, the theoretical issues of what constitute sustainability provide for lively debate, but the realization of the principles need, above all, pragmatic and practical means of expression in the field. Without indicators, sustainability will remain confusing and complex and hence prone only to rhetorical quotation.

Secondly, one overriding lesson from recent field experience is that the development of more sustainable systems can be planned only with the full involvement of communities who are expected to shoulder the risk of change. These communities must participate in the testing of options and take responsibility for the outcomes. This does not mean that knowledge gained from outside the local system is irrelevant. On the contrary, systems often evolve and develop from the injection of new ideas. The local emphasis on empowerment and peoples' decision-making implies a major switch in national policies and approach that will not always be easy to implement.

Thirdly, it has been shown that the application of sustainability issues can have a significant impact on landscapes and land use. For example, the Landcare Programme in Australia involves substantial co-operation between government agencies, academic institutions and resource users, and the programme has quite dramatically reordered personal priorities and actual on-farm activities. In Western Australia since 1988, the Department of Agriculture has embarked on a programme of facilitating group empowerment in the Landcare movement. Training in facilitation techniques, and management and leadership skills, is given (Hartley et al. 1992). A simplified relationship between empowerment and information demand, with the dual objectives of agricultural production and environmental sustainability, is given in Figure 2.1. So, a key question of practicality has been demonstrated and the concept is no longer in the realm of impractical theory.

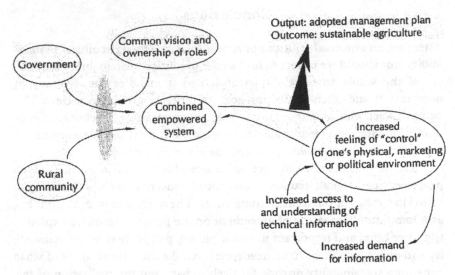

Figure 2.1 Relationship between empowerment and the demand for information: a model for extension in Western Australia's Landcare Movement (after Hartley et al. 1992).

Fourthly, the application of the concept of agricultural sustainability would not be acceptable as a challenge to important export crop incomes of many developing countries. Estate production is perceived as essential for maximizing production of crops such as oil palm, as well as guaranteeing output quality. One of the lasting legacies of colonialism is that production can be maintained only if large-scale, capital-intensive agriculture remains. To a degree, within the prevailing economies of developing countries, this is true. However, smallholder mixed farming can also be both productive in the sense of producing crops that are valuable for export and supportive of local livelihoods (the case of intensive cassava production for export to the Netherlands is one example, along with tree crops and rice for subsistence on the infertile soils of Thailand). At the same time, there have been significant changes in farm household income generation in many countries; food crops are important cash crops in local and international markets and off-farm income is becoming an important source of total household income. The assumption that rural dwellers are always farmers is breaking down: many are small-scale entrepreneurs, and the diversity of such activities is a major contributor to the sustainability of the local economy and hence of the land use itself.

Finally, lower external input systems can be both highly productive and sustainable. The evidence is, however, still largely based on incomplete and fragmentary empirical studies, and much more research is needed to examine the social, physical and biological basis for the potential for low external-input agriculture in a wide variety of environments and with many different types of farmers. Much of the critical reaction to low-input agriculture is based upon prejudice, and much of the support is nothing more than wishful thinking. It is vital that basic questions are asked about these systems, such as their replicability, long-term performance and ability to support changing life-styles of human populations.

Any move towards the design and implementation of sustainable agriculture not only involves farmer-led innovations but also institutional, policy and trading changes at many levels (Gibbon 1991). The realization of practical systems of land use may seem somewhat optimistic, considering the many conflicting political and macroeconomic influences extant today. For example, the uncertain outcomes of the recently concluded GATT agreements is a worry for poorer farmers in developing countries. Much of the better food security of India has been built upon protection for local agricultural commodities, and not the competition of international markets. However, as value systems change as a result of better knowledge about people–environment interactions, political attitudes will change and eventually be represented by effective policy changes. This has happened already for Green Revolution agriculture, and is happening now in some developed countries for some environmental issues related to agriculture such as nitrate pollution. Although the change may be slow, there is no doubt that there is a recognition by many that our current methods of utilizing natural resources are vulnerable. The pressure to adjust agriculture to more sustainable practices will become greater as the whole subject of sustainability moves up the political agenda. Meanwhile, are more sustainable agricultural systems possible? Most certainly, yes. Will we see such systems in our lifetimes? Maybe, in some places and by some groups. There will be much negotiation and re-education before the rhetoric of sustainability pervades all layers of agriculture from subsistence peasant farming to multinational agri-business.

References

Abel, N. O. J. & M. A. Stocking 1981. The experience of underdeveloped countries in the appraisal of development schemes and the review of policy. In *Progress in resource management and environmental planning*, T. O'Riordan & D. Sewell (eds), 253–95. Chichester: John Wiley.

Adams, W. M. 1990. *Green development: environment and sustainability in the Third World*. London: Routledge.

Alexandratos, N. (ed.) 1988. *World agriculture toward 2000: an FAO study*. London: Pinter (Belhaven).

Altieri, M. A. 1987. *Agroecology: the scientific basis of alternative agriculture*. London: IT Publications.

Baker, D. & D. Norman 1990. The farming systems research and extension approach to small farm development. In *Agroecology and small farm development*, M. A. Altieri & S. B. Hecht (eds), 91–101. Boca Raton, Louisiana: CRC Press.

Bartelmus, P. 1994. *Environment, growth and development: the concepts and strategies of sustainability*. London: Routledge.

Bawden, R. 1990. Of agricultural systems and systems agriculture: systems methodologies in agricultural education. Paper presented for a festschrift to mark Colin Spedding's retirement, Department of Agriculture, University of Reading.

Bennett, A. S. 1991. Aid to natural resources – a forward look. *Tropical Agricultural Association Newsletter* 11, 1–6.

Blaikie, P. & H. Brookfield (eds) 1987. *Land degradation and society*. London: Methuen.

Brink, B. ten 1991. The AMOEBA approach as a useful tool for establishing sustainable development? In *In search of indicators of sustainable development*, H. Verbruggen & O. Kuik (eds), 71–87. Dordrecht: Kluwer.

Brokensha, D., D. M. Warren, O. Werner 1980. *Indigenous knowledge systems and development*. Madison: University Press of America.

Brown, B. J., M. E. Hanson, D. M. Liverman, R. W. Meredith 1987. Global sustainability: toward definition. *Environmental Management* 11, 713–19.

Brown Jr, G. E. 1984. Stewardship in agriculture. In *Agricultural sustainability in a changing world order*, G. K. Douglass (ed.), 104–131. Boulder, Colorado: Westview.

Brundtland, G. H. (Chair) 1987. *Our common future* [The World Commission on Environment and Development]. Oxford: Oxford University Press.

Capra, F. 1983. *The turning point: science, society and the rising culture*. London: Flamingo Books.

Chambers, R., A. Pacey, L. A. Thrupp 1989. *Farmer first: farmer innovation in agricultural research*. London: Intermediate Technology Publications.

Checkland, P. B. 1988. Images of systems and the systems image. *Journal of Applied Systems Analysis* 15, 37–42.

Chissano, J. A. 1994. Natural resource management and the environment: widening the agricultural research agenda. Theme essay in ISNAR *Annual Report 1993*, International Service for National Agricultural Research, The Hague, 35–47.

Clunies Ross, T. & N. Hildyard 1992. *The politics of industrial agriculture*. London: Earthscan.

Colby, M. E. 1989. *The evolution of paradigms of environmental management in development*. Strategic Planning and Review Discussion Paper, World Bank, Washington DC.

Conroy, C. & M. Litvinoff 1988. *The greening of aid – sustainable livelihoods in practice*. London: Earthscan (in association with the International Institute for Environment and Development).

Conway, G. R. 1985. *Agroecosystem analysis for research and development*. Bangkok: Winrock International Institute for Agricultural Development.

— 1987. The properties of agroecosystems. *Agricultural Systems* 24, 95–117.

REFERENCES

— 1991. Sustainability in agricultural development: trade-offs with productivity, stability and equitability. Paper presented at the 11th annual AFSR/E Symposium, Michigan.

Cornwall, A., I. Guijt & A. Wellbourn 1992. Acknowledging process: challenges for agricultural research and experimental methodology. Overview paper for workshop, "Beyond Farmer First: Rural Peoples' Knowledge', International Institute for Environment and Development, and Institute of Development Studies, Sussex.

Dahlberg, K. A. 1984. Ethics, values and goals in agricultural systems and agricultural research. In *Sustainable agriculture and integrated farming systems,* T. C. Edens, C. Frigden, S. L. Battenfield (eds), 202–218. East Lansing: Michigan State University Press.

Douglass, G. K. (ed.) 1984. *Agricultural sustainability in a changing world order.* Boulder, Colorado: Westview.

Elliott, J. A. 1994. *An introduction to sustainable development.* London: Routledge.

FAO 1991. *The den Bosch declaration and agenda for Action on Sustainable Agriculture and Rural Development.* Report of the Conference, UN Food and Agriculture Organisation, Rome.

— 1992. *Sustainable development and the environment: FAO policies and actions, Stockholm 1972 – Rio 1992.* Rome: UN Food and Agriculture Organisation.

— 1993. *Agriculture: toward 2010.* Conference of the 27th Session of the UN Food and Agriculture Organisation. Rome: UN Food and Agriculture Organisation.

Farrington, J. & A. M. Martin 1987. Farmer participatory research: a review of concepts and practice. Discussion Paper 19, Agricultural Administration Unit, Overseas Development Institute, London.

Firebaugh, F. M. 1990. Sustainable agricultural systems: a concluding view. In *Sustainable agricultural systems,* C. A. Edwards, R. Lal, P. Madden, R. H. Miller, G. House (eds), 674–6. Ankeny, Iowa: Soil and Water Conservation Society.

Fukuoka, M. 1988. *The natural way of farming.* Tokyo: Japan Publications.

Gibbon, D. 1991. Farming systems research for sustainable agriculture: the need for institutional innovation, participation and iterative approaches. Paper presented at the seminar of the CERES Project on "Farm Household Strategies", Vila Real, Portugal.

Giller, K. E., J. F. McDonagh, G. Cadisch 1994. Can biological nitrogen fixation sustain agriculture in the tropics? In *Soil science and sustainable land management in the tropics,* J. K. Syers & D. L. Rimmer (eds), 173–91. Wallingford: CAB International.

Goodland, R., H. Daly, S. El-Serafy, B. von Droste (eds) 1991. *Environmentally sustainable economic development – building on Brundtland.* Paris: UNESCO.

Greenland, D. J. 1994. Soil science and sustainable land management. In *Soil science and sustainable land management in the tropics,* J. K. Syers & D. L. Rimmer (eds), 1–15. Wallingford: CAB International.

Grubb, M. 1993. *The Earth Summit agreements: a guide and assessment.* London: Earthscan.

Hanson, A. J. & G. Regallet 1992. Communicating for sustainable development. *Nature and Resources* 28, 35–43.

Harrington, L. W. 1991. Measuring sustainability: issues and alternatives. Paper presented at the 11th annual AFSR/E Symposium, Michigan, 5–10th October.

Harrington, L. W. & P. Hobbs 1992. Assessing the sustainability of the rice–wheat

65

cropping pattern through farmer monitoring: an example from the Nepal Terai. Presented at the Second Asian Farming Systems Symposium, Colombo, Sri Lanka, 2–5 November.

Hartley, R. E. R., J. R. H. Riches, J. K. Davis 1992. A systems approach to Landcare. In *People protecting their land* (proceedings of the 7th ISCO conference), P. G. Haskins & B. M. Murphy (eds), 217–22. Sydney: Department of Conservation and Land Management.

Haverkort, B., J. van der Kamp, A. Waters-Bayer 1992. *Joining farmers' experiments – experiences in participatory technology development*. London: Intermediate Technology Publications.

Haverkort, B., D. Gibbon, W. Bayer 1991. *Research for the future: an overview of existing research in sub-Saharan Africa for the development of low external-input and sustainable agriculture*. Leusden, The Netherlands: SAREC/Information Centre for Low External Input Agriculture.

ILEIA 1988. Towards sustainable agriculture: abstracts, periodicals, organizations. May 1988. Leusden, Netherlands.

IUCN 1980. *World conservation strategy: living resource conservation for sustainable development*. Gland, Switzerland: International Union for Conservation of Nature and Natural Resources, United Nations Environment Programme and World Wildlife Fund.

IUCN/UNEP/WWF 1991. *Caring for the Earth: a strategy for sustainable living*. Gland, Switzerland: The World Conservation Union (IUCN), United Nations Environment Programme and the World Wide Fund for Nature.

Jodha, N. S. 1990. Sustainable agriculture in fragile resource zones: technological imperatives. Mountain Farming Systems. Discussion Paper 3, ICIMOD, Kathmandu, Nepal.

Lockeretz, W. 1988. Open questions in sustainable agriculture. *American Journal of Alternative Agriculture* 3, 174–81.

Lundgren, L., G. Taylor, A. Ingevall 1993. *From soil conservation to land husbandry: guidelines based on SIDA's experience*. Stockholm: Swedish International Development Authority.

Meadows, D. H., D. L. Meadows, J. Randers, W. Behrens III 1972. *The limits to growth: a report for the Club of Rome's project for the predicament of mankind*. London: Pan.

Meadows, D. H., D. L. Meadows, J. Randers 1992. *Beyond the limits – global collapse, or a sustainable future*. London: Earthscan.

Mollison, B. 1988. *Permaculture: a designers' manual*. Tyalgum: Tagari Publications.

Murdoch, J. 1993. Sustainable rural development: towards a research agenda. *Geoforum* 24, 225–41.

NRC 1989. *Alternative agriculture*. Committee on the role of alternative farming methods in modern production agriculture. Board on Agriculture, National Research Council. Washington DC: National Academy Press.

Odum, H. T. 1959 *Fundamentals of ecology*. Philadelphia: W. B. Saunders.

Okali, C., J. Sumberg, J. Farrington 1994. *Farmer participatory research: rhetoric and reality*. London: Intermediate Technology Publications.

Okigbo, B. N. 1989. *Development of sustainable agricultural production systems in Africa: roles of international agricultural research centres and national agricultural research systems*. Ibadan, Nigeria: International Institute for Tropical Agriculture.

REFERENCES

O'Riordan, T. 1988. *The politics of sustainability*. In *Sustainable environmental management: principles and practice*, K. Turner (ed.), 29–50. London: Pinter (Belhaven).

— 1995. The global environment debate. In *Environmental science for environmental management*, T. O'Riordan (ed.), 16–29. Harlow: Longman.

Pearce, D. W. & J. J. Warford 1993. *World without end: economics, environment and sustainable development*. Oxford: Oxford University Press.

Pimentel, D. 1984. Energy flows in food systems. In *Food and energy resources*, D. Pimentel & C. Hall (eds), 1–23. New York: Academic Press.

Ploeg, van der J. 1990. *Labor, markets and agricultural production*. Boulder, Colorado: Westview.

Premkumar, P. D. 1994. *Farmers are engineers: participative and integrated development of watershed project, Gulbarga, Karnataka, India*. Karnataka: MYRDA.

Pretty, J. & R. Sandbrook 1991. *Operationalising sustainable development at the community level: primary environmental care*. London: International Institute for Environment and Development.

Redclift, M. 1987. *Sustainable development: exploring the contradictions*. London: Methuen.

— 1991. The multiple dimensions of sustainable development. *Geography* 76, 36–42.

Redclift, M. & C. Sage (eds) 1994. *Strategies for sustainable development: local agendas for the Southern Hemisphere*. Chichester: John Wiley.

Reijntjes, C., B. Haverkort, A. Waters-Bayer 1992. *Farming for the future: an introduction to low external input agriculture*. London: Macmillan.

Remigio, A. 1994. Recent Philippine political economy and the sustainable development paradigm. In *Strategies for sustainable development: local agendas for the Southern Hemisphere*, M. Redclift & C. Sage (eds), 61–95. Chichester: John Wiley.

Richards, P. 1985. *Indigenous agricultural revolution*. London: Hutchinson.

Rodale, R. 1983. Breaking new ground: the search for sustainable agriculture. *The Futurist* 1, 15–20.

Shaner, W. W., P. F. Philip, W. R. Schmel 1982. *Farming systems research and development: guidelines for developing countries*. Washington DC: World Bank.

Sousan, J. G. 1992. Sustainable development. In *Environmental issues in the 1990s*, A. M. Mannion & S. R. Bowlby (eds), 21–36. Chichester: John Wiley.

Steiner, R. 1974. *Agriculture*. London: Biodynamic Agricultural Association

Stocking, M. A. (ed.) 1994. *Integrating environment and sustainable development themes into agricultural education and extension programmes: report of an expert consultation*. Rome: Human Institutions and Agrarian Reform Division, UN Food and Agriculture Organisation.

— 1995. Soil erosion and land degradation. In *Environmental science for environmental management*, T. O'Riordan (ed.), 223–42. Harlow: Longman.

Swift, M. J., B. T. Kang, K. Mulongoy, P. W. Woomer 1991. Organic matter management for sustainable soil fertility in tropical cropping systems. In *Evaluation for sustainable land management in the developing world*, J. Dumanski, E. Pushparajah, M. Latham, R. Myers (eds), 307–326. IBSRAM Proceedings 12 (vol. 21), Bangkok, Thailand.

Tiffen, M., M. Mortimore, F. Gichuki 1994. *More people, less erosion*. Chichester: John Wiley.

Turner, R. K. 1988. *Sustainable environmental management*. London: Pinter (Belhaven).

Verbruggen, H. & O. Kuik (eds) 1991. *In search of indicators of sustainable development*. Dordrecht: Kluwer Academic Press.

WCED (World Commission on Environment and Development) 1987. *Our common future*. Oxford: Oxford University Press.

Wiseman, H., J. Vanderkop, J. Nef 1991. *Critical choices! Ethics, science and technology*. Toronto: Thompson Educational.

CHAPTER THREE

Global warming and development

Mick Kelly & Sarah Granich

If you have come to help me you can go home again. But if you
see my struggle as part of your own survival then perhaps we can
work together.   Australian Aborigine Woman

Editors' introduction

In the past decade there has been a major debate as to whether human-
induced global warming is real or just an illusion. The overall consensus is
that it is real, but scientists differ in their predictions as to by how much the
Earth will warm over the coming years. Sophisticated computer models that
try to predict the temperature of the Earth are beginning to provide insights
into the likely effects of releasing pollutants such as carbon dioxide and sul-
phate particles into the atmosphere (Mathews 1994). However, even if the
predictions of such models turn out to be wrong, the risk of not taking action
now is immense. Indeed, it has been convincingly argued that concerns over
the global environmental rather than development was the major force that
spawned and directed the Earth Summit held in Rio during 1992 (Middleton
at al. 1993). However, is it the case that such environmental concerns only
become "global" once they threaten quality of life in the rich North?

Global climatic change received a great deal of attention at the Rio summit. The Framework Convention on Climate Change signed during the summit required that governments seek to achieve (Article 2): "stabilisation of
greenhouse gas concentrations in the atmosphere at a level that would prevent dangerous anthropogenic interference with the climate system".

The very fact that the Climate Convention was signed by many governments, including those from the rich North, was a significant achievement of
the Rio summit, although the often vague language and commitments specified in the document have been rightly criticized (Middleton at al. 1993).

Whether these commitments are enough to avoid some of the potential impacts of global warming is another question.

In this chapter, Mick Kelly & Sarah Granich explore some of the potential impacts that global warming may have on developing countries. As the relationships between developed and developing countries are complex, then predicting some of these impacts is fraught with difficulties. Therefore, to some extent this chapter, like the one on biotechnology (Ch. 5), is speculative, but both chapters nonetheless illustrate the need for considered thought in areas where there could be potentially massive upheaval.

Global warming is also an example of a scientific "fact" that is played out in the arena of political ecology, and subject to the interpretation and pressures of many different interest groups. Conflicts arose at Rio between the North, which saw the conservation of forests as an essential buffer against global warming, and the South, which put a greater emphasis on development. Piers Blaikie explores this arena of political ecology more fully in Chapter 1, and Stocking at al. Dissect the interaction between conservation and development in Chapter 6. A gender perspective on environmental change is provided by Jackson (Ch. 4).

Introduction

Global warming is the term used to describe the climate change induced by increasing concentrations of "greenhouse" gases in the atmosphere. Gases such as carbon dioxide, methane, nitrous oxide and the halocarbons trap heat near the Earth's surface. As emissions of these gases continue to rise, it is feared that the changing composition of the atmosphere will result in a substantial change in regional climates and a rise in sea level, with considerable impacts on human wellbeing. Because greenhouse gas emissions are the result of activities fundamental to the traditional process of development (industrialization, rising energy use and expanding agriculture) emission control presents a serious challenge to conventional views of economic growth. Adaptation to the changing environment may also prove a difficult goal to achieve, particularly for the developing nations who are at most risk from climate change and sea level rise.

Recognizing that climate change is a global problem that warrants a global response, 155 nations signed the Framework Convention on Climate Change at the United Nations Conference on Environment and Development in June 1992. The Convention is intended to reduce greenhouse gas emissions and to mitigate the most severe consequences of global warming.

It represents the culmination of four intense years of scientific review and political negotiations and it will shape the international response to the threat of global warming over years to come. In calling attention to the need to integrate climate concerns in all aspects of the development process, whether in the North or in the South, it will also exert an important influence on the development agenda. In highlighting the importance of considering long-term environmental change, climate change provides a valuable long-term perspective from which to view the processes that may lead to sustainable development.

This chapter considers the relevance of global warming to the development process and the nations of the Third World, placing the discussion in the context of the Framework Convention on Climate Change. We first describe the historical process that led to the Framework Convention, resulting in a ground-breaking climate treaty a mere five years after the world was alerted to the problem of global warming. We then look at the reasons why the international community is concerned about climate change, considering the specific situation of the developing nations. Within its text, the Framework Convention contains reference to many aspects of the debate concerning the response to climate change, and we discuss various issues that surfaced during the treaty negotiations. Finally, we consider how the aims of the Convention may be put into practice, drawing out the perspective on sustainable development that climate change affords.

Gribbin & Kelly (1989), Schneider (1989) and Gribbin (1990) provide comprehensive introductions to the issue of global warming for the general reader, and Foley (1991) considers aspects of particular relevance to developing nations. A more technical introduction can be found in Leggett (1990). An excellent interpretation of the current state of understanding of the problem, directed towards those concerned about policy responses, can be found in Mintzer (1992).

When temperatures rose beyond one hundred degrees in many cities in the United States during the summer of 1988, drought and warmer climates were no longer ignored as merely "local" occurrences in the Third World. Climatic change had become a "global" problem. The millions of deaths in dozens of countries did not make the tragedy global, because it took place in the Third World. It remained local. Thermometers registering a few degrees more in the United States, however, succeeded in turning climatic change into a global issue for all the governments of the industrialized north and the entire scientific community were immediately mobilized. Scientific uncertainty, which had been used as an excuse for inaction on burning problems of the Third World, had suddenly become acceptable in policy-making, because the issues now touched the privileged North. Vandana Shiva, Research Foundation for Science and Ecology

Historical perspective

As long ago as the 1890s, the Swedish scientist Svante Arrhenius forecast that coal burning could release sufficient carbon into the atmosphere to change global climate. His warning was largely ignored. Scientific interest in the greenhouse effect waxed and waned during the first half of the twentieth century, yet the issue received little attention outside the scientific community. Current concern about global warming has its roots in the dramatic improvement in scientific understanding of the problem that has taken place over the past two decades (Gribbin & Kelly 1989). During the early 1980s, a series of authoritative reviews documented the scientific consensus that was gradually emerging. Two major international studies, one undertaken by the International Council of Scientific Unions (ICSU), UNEP and the World Meteorological Organization (WMO) and the other by the US Department of Energy, clearly signalled the scientific community's concern about global warming. The ICSU/UNEP/WMO authors concluded that climate change represents "one of today's most important long-term environmental problems", sounding the warning bell with typical academic understatement.

In the late 1980s, a spate of climatic disasters focused the world's attention on the impact of extreme weather and climate. In 1987, monsoon failure in India resulted in an 85 million tonne fall in world grain production. With a poor harvest in China, global production dropped by another 76 million tonnes leaving world grain reserves at their lowest level for 15 years. The following year, flooding killed more than a thousand and left 20 million homeless in Bangladesh. During the closing months of that year, harvest failure in Ethiopia threatened over 4 million people, raising once again the appalling prospect of mass starvation. However, the event that finally triggered widespread concern about global warming was the drought that affected the Midwest of the USA during the summer of 1988. It is a sad comment on human nature that it took drought in the American heartland to attract the attention of politicians and others, whereas for decades millions have gone hungry elsewhere in the world. Nevertheless, drought in the Midwest continues to be a cause for alarm for not only local farmers and consumers but also the many countries around the world dependent on surplus production from that region.

There has been a growing realization . . . that it is impossible to separate economic development issues from environment issues; many forms of development erode the environmental resources upon which they must be based, and environmental degradation can undermine economic development. Poverty is a major cause and effect of global environmental problems. It is therefore futile to attempt to deal with environmental problems without a broader perspective that encompasses the factors underlying world poverty and international inequality.   *Our common future*

Humanity is conducting an unintended, uncontrolled, globally pervasive experiment whose ultimate consequences could be second only to a global nuclear war. The Earth's atmosphere is being changed at an unprecedented rate by pollutants resulting from human activities, inefficient and wasteful fossil fuel use and the effects of rapid population growth in many regions. These changes represent a major threat to international security and are already having harmful consequences over many parts of the globe . . . It is imperative to act now.
The Changing Atmosphere: Implications for Global Security, Toronto, June 1988

The conference, The Changing Atmosphere: Implications for Global Security, was held in Toronto in June 1988. Against the backdrop of the Midwest drought, scientists, environmentalists, politicians and decision-makers met to consider the broad threat posed by atmospheric pollution. It was at this point that global warming was transformed from a subject of largely academic concern to an issue ranking high on the political agenda. It is not surprising, given the vulnerability of the Third World, that one of the first reports inspired by this surge in concern about the issue was undertaken at the request of the Commonwealth Heads of Government, meeting in 1987.*Meeting the challenge* (Anon 1989) presents a comprehensive overview of the implications of global warming and an action plan for the Commonwealth nations. The authors note that "new and bold policies, backed by more resources than hitherto, are expected by the governments and peoples of the world" to deal with global warming.

Sir Shridath "Sonny" Ramphal, for 14 years Secretary General of the Commonwealth, warned that:

> there is no way out of the environment problem and the mess we are in without a positive approach to development . . . We have got to understand that we are not just, as we were centuries ago, or maybe a decade ago, a world of many States. We have got to think of the state

of our one world . . . The alternative is that our common future is in jeopardy.

Global warming provides a new imperative for sustainable development, providing a graphic example of the dangerous consequences of current models of economic growth. As concern about global warming grew, so too did awareness that humanity faces environmental catastrophe on many fronts. The impact of the thoughtless exploitation of resources had become too obvious to ignore. *Our common future*, the report of the World Commission on Environment and Development, was published in 1987 (WCED 1987). The most significant aspect of *Our common future* was the stress it placed on the fundamental link between the state of the environment, social justice and the nature of development.

> The Convention was negotiated in record time. From the beginning of official negotiations in Washington to adoption and thereafter signature was a little over one year. There were only 61 days of official negotiations. Although some may argue it took a long time for the international community to act on the issue of human-induced climate instability, the international legal process cannot be criticized for being too slow. All those that participated in the Intergovernmental Negotiating Committee (INC) for the Climate Change Convention – and more parties, both government and non-governmental, participated in this agreement than in any other in history – will remember the extraordinary experience of consensus-building in the UN Chamber . . . The negotiators understood all too well the difficulty of overcoming vested interests in the energy sector. However, I would venture to say that even their own negotiators were at least impressed by the overwhelming desire for change. Change not for the sake of it but for the sake of returning to a form of balance and stability in the ecosystem, which will enable our generation to feel confident that we can pass the systems that support life on Earth to our children and their children in a healthy state.   James Cameron, Foundation for International Environmental Law and Development

Two agencies took the lead in co-ordinating the international response to global warming. The World Meteorological Organization(WMO) is responsible for co-operation between national meteorological services worldwide. The brief of the United Nations Environment Programme (UNEP) is to identify environmental problems, monitor scientific developments and promote and guide the international response. In the closing years of the decade, UNEP and the WMO organized the Intergovernmental Panel on Climate Change (IPCC). The aim of the IPCC, whose work continues, is to provide an up-to-date and authoritative assessment of scientific understanding of the problem and of the implications of climate change that will guide the international community in its response. The IPCC issued its first series of reports in August 1990, concluding that, although scientific uncertainties remained, prompt action was needed to combat the threat of climate change. This recommendation was endorsed by the Ministerial Declaration of the Second World Climate Conference held in Geneva in late 1990 (Jager & Ferguson 1991). Shortly afterwards, the United Nations General Assembly called for an international treaty on climate change. The Intergovernmental Negotiating Committee for a Framework Convention on Climate Change (INC) was then established to draft the text of the climate treaty. It was at this point that what had been a largely scientific and technical process became heavily politicized. With a climate convention a tangible prospect, national delegates established hard negotiating positions, and the gulf between the common good and national self-interest became ever more apparent. Nevertheless, the text of the Framework Convention on Climate Change was agreed in time for signature at the Earth Summit in Rio de Janeiro in June 1992. Despite criticism of the many compromises and of the vagueness and ambiguity of the Convention, this must be considered a major achievement. The Convention came into force on 21 March 1994.

- Climate change is a global issue; effective responses would require a global effort that may have a considerable impact on humankind and individual societies.
- Industrialized countries and developing countries have a common responsibility in dealing with problems arising from climate change.
- Industrialized countries have specific responsibilities on two levels:
  - a major part of emissions affecting the atmosphere at present originates in industrialized countries where the scope for change is greatest. Industrialized countries should adopt domestic measures to limit climate change by adapting their own economies in line with future agreements to limit emissions;
  - to co-operate with developing countries in international action, without standing in the way of the latter's development, by contributing additional financial resources, by appropriate transfer of technology, by engaging in close co-operation concerning scientific observation, by analysis and research, and finally by means of technical co-operation geared to forestalling and managing environmental problems.
- Emissions from developing countries are growing and may need to grow in order to meet their development requirements and thus, over time, are likely to represent an increasingly significant percentage of

global emissions. Developing countries have the responsibility, within the limits feasible, to take measures to suitably adapt their economies.

- Sustainable development requires the proper concern for environmental protection as the necessary basis for continuing economic growth. Continuing economic development will increasingly have to take into account the issue of climate change. It is imperative that the right balance between economic and environmental objectives be struck.

- Limitation and adaptation strategies must be considered as an integrated package and should complement each other to minimize net costs. Strategies that limit greenhouse gas emissions also make it easier to adapt to climate change.

- The potentially serious consequences of climate change on the global environment give sufficient reasons to begin by adopting response strategies that can be justified immediately even in the face of significant uncertainties.

- A well informed population is essential to promote awareness of the issues and provide guidance on positive practices. The social, economic, and cultural diversity of nations will require tailored approaches.   IPCC Response Strategies Report

One of the most striking developments during the INC negotiations was the formation of a series of Third World alliances. The G77 Group of developing nations maintained a common position through most of the discussions, and several other regional alliances sprang into existence. The most notable and vociferous was the Alliance of Small Island States (AOSIS), representing the nations most vulnerable to, and least able to cope with, sea level rise. AOSIS has proposed an innovative "insurance pool" to compensate vulnerable nations for the adverse effects of sea-level rise (Anon. 1992). It was the lobbying of these Southern groups that led to what may prove to be an important aspect of the Framework Convention: the section on principles, unique in this form of international agreement. The degree to which nongovernmental organizations participated in the INC process was also notable and it set an important precedent for future environmental conventions.

The Framework Convention on Climate Change is a statement of aims and goals, containing some specific commitments but lacking in detail. It was designed this way to ensure the agreement of the broadest possible range of countries and it contains some deliberate ambiguities to maximize adoption. The next stage in the process is the negotiation of a series of detailed protocols covering specific issues, such as emission control targets for particular gases. Negotiations began in December 1992 and are likely to continue through the decade as the many different aspects of the problem are dealt with one by one.

The scientific basis of concern

The climate system encompasses the Earth's atmosphere, oceans, cryosphere and biosphere. Determining the influence of human activity on the system and then the impact of climate change on humanity requires detailed understanding of social, economic and political processes. Because of the complexity of the problem, predicting the ultimate impact of global warming is not an easy matter, stretching scientific understanding to the limit. Forecasts are, by their nature, speculative. They rest on uncertain projections of social change, highly simplified computer models of physical, biological and social processes, experience and conjecture. Nevertheless, sufficient information is available to support the conclusion that global warming may pose a grave threat to human welfare. As discussed below, the scientific basis of concern rests on four pillars. Readers interested in further details of the scientific case are directed to Leggett (Leggett 1990), the IPCC science reports (Houghton et al. 1990, Houghton et al. 1992) and Mintzer (1992).

The greenhouse effect

Many of the pollutants released into the atmosphere as a result of human activity – such as carbon dioxide, the chlorofluorocarbons (CFCs), methane and nitrous oxide – are greenhouse gases and generate what is known as "the greenhouse effect". The greenhouse effect itself is a well established physical phenomenon. If it were not for the greenhouse gases found naturally in the atmosphere, such as carbon dioxide and water vapour, the planet would be some 30°C colder than it is at present. These gases allow energy from the Sun to pass down through the atmosphere unhindered, but trap heat radiated by the surface of the planet that would otherwise escape directly to space. Global warming is a problem because humanity is releasing greenhouse gases into the atmosphere at an ever-increasing rate, intensifying the greenhouse effect.

The changing composition of the atmosphere

There is no doubt that the composition of the atmosphere has undergone a radical change as a result of industrialization. The amount of carbon dioxide in the air has risen by a quarter since the late eighteenth century, to the highest point for 160 000 years, and the rate of change is accelerating; about half this increase has occurred since the 1950s. There is now twice as much methane in the air as there was three centuries ago, and levels of the other major greenhouse gases – the chlorofluorocarbons, nitrous oxide and surface ozone – are all rising significantly. Although uncertainties remain concerning the detail, the overarching cause of this change in atmospheric composition is clear. It is the result of human activity. The major contributors are shown in Figure 3.1: the growth in energy consumption, the products of industrialization and accelerating land-use change. That these activities are fundamental to the historical process of development is a key characteristic of the global warming problem.

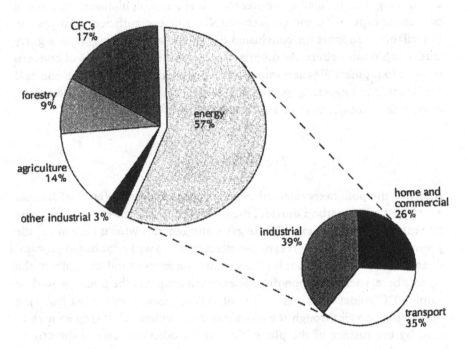

Figure 3.1 Sources of greenhouse gases by sector during the 1980s (*source:* Lashof & Tirpak 1989).

Global warming has already occurred

The change in atmospheric composition over recent centuries should already have affected global climate. Computer models predict that global warming of 0.3–1.1°C should have occurred over the past hundred years. The world has, in fact, warmed by about 0.5°C during this period. The surface temperature of the planet is now higher than it has been at any time during the period of instrumental observations, which began in the mid-

nineteenth century. That there is broad agreement between the model forecasts and real-world experience provides strong support for the greenhouse
theory, but it is not conclusive. There are other possible, if less plausible,
explanations for the warming trend. The temperature rise must continue for
some years, surpassing the natural variability in climate, before other explanations can be wholly discounted and the greenhouse effect firmly established as the cause.

The future impact

Current projections, based on estimates of emission trends and model sim-
ulation of effects on climate, suggest that by the year 2050 global temperature may stand up to 3°C higher than now (Fig. 3.2). Placing this in
historical context, the temperature rise that brought the planet out of the
most recent ice age was only about 4°C. That change in climate took thousands of years, not decades. The world has not experienced temperature
levels 2°C higher than at present for 125 000 years. There is, though, considerable uncertainty in the projections of future climate change. For example, uncertainty as to how the climate system as a whole will respond to a
strengthening of the greenhouse effect results in the threefold range in the
temperature projections (Fig. 3.2). As temperatures rise, the amount of
water vapour in the atmosphere will change, the distribution of cloud will
alter, the oceans will respond, and the snow and ice margin will shift. Factors
such as these will amplify or reduce the initial temperature response, producing complex "feedback" effects, and it is impossible at this time to determine how weak or strong these factors are. It may be decades before this
uncertainty is resolved. Nevertheless, consideration of the likely impact of
these changes shows that global warming represents a substantial change in
the planetary environment, even if the more optimistic forecasts are correct.

The impact of global warming

Even today, many aspects of human activity, from forestry and the management of nature reserves to agriculture and energy production, are sensitive
to periods of extreme weather and climate. As global warming progresses,
four aspects of the change in climate are likely to generate the most wide-
ranging impacts.

- The speed of change may be too great to allow the natural migration

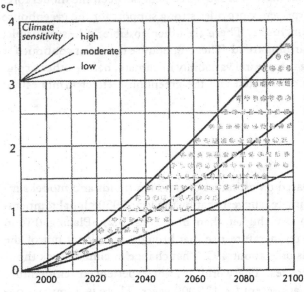

°C

Figure 3.2 A "business as usual" projection of the response of global-mean surface air temperature to greenhouse gas emissions. The projection assumes no concerted effort to curb emissions beyond the halocarbon controls under the Montreal Protocol and subsequent amendments. The shaded area indicates the range of uncertainty. (*Source*: Wigley & Raper 1992.)

of species, resulting in extinctions or substantial change in ecosystem characteristics and a failure of the support services they provide to humanity.

- As temperatures rise, the oceans will warm and sea water will expand. Glaciers on land will melt, resulting in a rise in sea level – perhaps by as much as one metre by the second half of the twenty-first century.
- Food production is likely to be affected directly by climate change and sea-level rise in many parts of the Third World.
- It is likely that the northern continental interiors will dry out as a result of the combined effects of rising temperature and decreased rainfall, with major implications for crop yields in northern middle latitudes, and worldwide repercussions, as surplus production is lost.

Biological diversity, already being reduced by various human activities, may be one of the chief casualties of global warming. Massive destruction of forests, wetlands, and even the polar tundra could irrevocably destroy complex ecosystems that have existed for millennia. Indeed, various biological reserves created in the past decade to protect species diversity could become virtual death traps as wildlife attempt to survive in conditions to which they are poorly suited. Accelerated species extinction is an inevitable consequence of a rapid warming.   Lester Brown, Christopher Flavin and Sandra Postel, Worldwatch Institute.

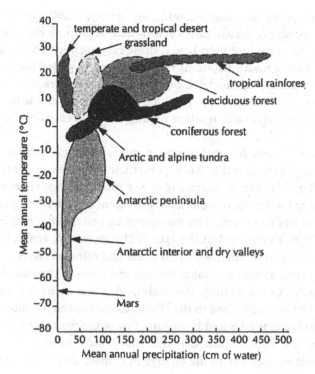

Figure 3.3 The sensitivity of major habitats to temperature and rainfall (*Source:* Harwell & Hutchinson 1985).

The IPCC impact report (Tegart et al. 1990) reviews current understanding of the impact of global warming, covering potential consequences for agriculture, land use, forestry, natural ecosystems, water resources and the world's oceans and ice cover, and it provides a pathway into the wider literature. Carter et al. (1992) summarize methods of assessing the impact of climate change and sea-level rise. See also relevant sections of Mintzer (1992).

Ecosystems in peril

Environmental conditions determine both the location and the productivity of the world's ecosystems. Many people, particularly in the Third World, depend directly on natural ecosystems for food, shelter and fuel. As climate changes, these resources will come under increasing pressure. Figure 3.3 shows how sensitive habitats are to temperature and rainfall. Some ecosystems can flourish only within a restricted temperature range, yet can tolerate a wide variation in water availability. The tropical forests, for example, are found only where the annual temperature lies between 20°C and 26°C, although the amount of rainfall can vary greatly. In contrast, the grasslands of middle latitudes can tolerate a wide range of temperature (3–30°C), but

81

are confined to areas where the annual rainfall lies between 250 mm and 1000 mm. Although there are many other factors that determine location and viability, such as soil type and light levels, these relationships enable a first estimate to be made of how ecosystems might respond to global warming. Tolerance levels will first be exceeded on the margins of current habitats, and ecosystems will become less productive. Then, as conditions worsen, some ecosystems will have to adapt or migrate or eventually they will become extinct.

However, will there be time for ecosystems to migrate to new habitats? The rate of change may be so great that it is beyond their adaptive capability. For example, forest boundaries can migrate only at a rate of tens of kilometres a century, yet global warming represents a movement of climatic zones of hundreds of kilometres a century. The mangrove can adapt to a rise in sea level of up to 80 mm a century, but the rate of change could, again, be ten times as great. Weakened by environmental stress and often bordered by impassable obstacles such as hills and lakes (towns and cities in the case of managed ecosystems such as cropland), the likelihood is that many species and ecosystems will be lost. According to the World Conservation Monitoring Centre in Cambridge in the United Kingdom, few of the world's nature reserves extend over the 200–300 km distance that would be needed if species were to migrate in response to the change in climate that is likely to occur by the middle of the twenty-first century.

The predicted effects of the change are unnerving: There will be significant shoreline movement and loss of land. A higher mean sea level would inevitably . . . increase the frequency of inundation and exacerbate flood damage. It would inundate fertile deltas, causing loss of productive agricultural land and vegetation, and increase saline encroachment into aquifers, rivers and estuaries. The increased costs of reconstruction, rehabilitation and strengthening of coastal defence systems could turn out to be crippling for most affected nations.

Maumoon Abdul Gayoom, President of the Maldives

Rising sea level

Sea-level rise represents a very direct threat to the three-quarters of the world's population living in low-lying areas close to the coast. Many major cities, port facilities and industries, and much fertile agricultural land, are located in coastal zones, so any impact will have widespread social and economic repercussions.

The smaller islands of the world's oceans are particularly vulnerable.

President Maumoon Abdul Gayoom of the Maldives has repeatedly stressed that, unless action is taken to curb global warming, his nation could literally disappear beneath the waves. The Republic of the Maldives is a nation of more than a thousand tiny islands at no point rising more than 3.5 m above sea level. Lying just over 1 m above sea level, much of the capital island, Male, including the international airport, was flooded by high seas in 1987. Inundation threatens many other small islands. According to research undertaken for the United Nations Environment Programme (UNEP), sea level rise would deal a "death blow" to Tokelau, Tuvalu, the Marshall Islands, the Line Islands and Kiribati in the Pacific, resulting in a flood of refugees. Nuku'alofa, home to a fifth of Tonga's population, would lose almost half its area if sea level rose by 1.5 m (Anon 1989).

Delta regions are also high-risk areas. Many of these are major centres of population, already prone to flooding. John Milliman and his colleagues, at the Woods Hole Oceanographic Institute in Massachusetts, have undertaken a comprehensive study of the implications of sea-level rise for the Nile and Ganges Deltas (Milliman et al. 1988). These fertile agricultural areas support a total of some 46 million people. Taking into account compounding factors such as the loss of sediment and subsidence, as well as the geological factors that influence local sea level, the researchers examined the impact of a series of sea-level rise scenarios. The Ganges Delta, spanning 650 km of coastline along the Bay of Bengal, has few defences. In this region, the worst case scenario – a rise in local sea level of 2 m by the year 2050 – results in the loss of 18 per cent of the habitable land of the region, with 15 per cent of the population displaced and 13 per cent of the gross national product lost. The Woods Hole study highlights the role that development policy will play in determining the final impact of global warming. In Bangladesh, groundwater extraction may already have doubled the rate of natural subsidence. In Egypt, nearly all sediment transport down the Nile has been halted by damming, increasing the delta's susceptibility to sea-level rise.

In the semi-arid tropics, which occupy 13 per cent of the world's land surface, production is finely tuned to what rainfall there is. Small variations in rainfall can produce major changes in production. Furthermore, the resources of farmers there are limited, and easily overstretched. Even a few years of drought can have dramatic consequences, as was the case during the African famine of the early 1980s. In the tropics, food production is related mainly to the monsoon. Any alteration in its timing or severity can also have a major effect on production. In both the tropics and the subtropics, agriculture is all the more vulnerable where carried out on marginal land – as is increasingly common.   Robin Clarke

Low-lying coasts have also been made more vulnerable by human activity. As the Commonwealth Secretariat group of experts has observed, "many low-lying tropical coasts are protected by coral reefs and mangroves. Both are under pressure in many areas from human activities including pollution, sedimentation as a result of bad land use and construction processes, dynamite fishing and coral block quarrying, and the excessive cutting of mangrove for poles and fuelwood" (Anon 1989). Ninety per cent of the population of Guyana, on the northeast coast of South America, live on the coastal plain, below high tide level and liable to inundation, erosion and flooding caused by runoff from the hills inland. A mere half-metre rise would place most of the population below sea level.

Dry statistics obscure the direct threat to life itself posed by storm-induced flooding. Higher sea level will mean that storm surges penetrate farther inland, a situation further aggravated if tropical storms occur with greater frequency and violence. Three hundred thousand people drowned in a single flood in East Pakistan, now Bangladesh, in 1970. A third of the country was inundated. It has proved impossible to protect the population of that region from flooding in the present day. How will they cope as global warming makes the risk even greater?

Agricultural pressure

In a world where food shortages and famine are already responsible for intolerable suffering, the implications of any further pressure on food production can only be considered catastrophic. Global warming might affect agriculture in several ways. Not only may the productivity and stability of harvests be affected, but also the regions in which particular crops can be grown. Sea-level rise may result in the loss of valuable agricultural land. Additional carbon dioxide in the atmosphere could boost the growth of certain crops and weeds, and pests may become more resistant to control as climate alters. Land management techniques could be affected by secondary environmental consequences, such as soil erosion. Planting and harvesting dates could alter drastically. Different crops and markets will have to be developed. New agricultural techniques will have to be adopted. It may be possible to adapt agricultural systems to the ever-changing environment. In some cases, if climate stabilizes, the new environment may be more favourable for food production. However, the process of adaptation will require considerable resources and it will take time. The time of transition will inevitably be a time of disruption.

> Governments in the North are going to look at their own national interests. There may be alliances that may be formed in that regard, but they are going to be alliances of those who live in the developed world, who want access to raw materials and want access to resources in order to continue to live in the style to which they have come accustomed. You will see the military establishment beginning to look at the Third World from a military standpoint.   Bill Arkin, Institute of Policy Studies

In a joint project sponsored by the International Institute of Applied Systems Analysis (IIASA, Laxenburg, Austria) and UNEP, teams of researchers have studied the effect of climate change on food production in a range of marginal environments (Parry et al. 1988). Their study of the impact of drought in semi-arid regions highlights the vulnerability of food production. In one case study, Raul Bravo (of the Ministry of Agriculture in Quito) and his team recorded the far-reaching consequences of the Ecuadorian drought that lasted from 1976 to 1980. Impacts were felt throughout the country, affecting livestock as well as cereal crops. Barley production dropped by over 60 per cent, the wheat harvest by almost a half. The consumer price index in Quito rose by a quarter. Food aid from the Food and Agriculture Organisation (FAO) increased tenfold by 1979. Based on their study, the authors concluded that the best response to drought would be additional government support for traditional coping techniques. In the longer term, improvement in the opportunities available to farmers – higher educational standards, health care, infrastructure development and so on – could make a significant difference.

As the change in climate progresses, limits of cropping areas may shift as the production of particular crops becomes no longer viable. Basing their assessment on the ratio of annual rainfall to potential evaporation (i.e. moisture availability), the IIASA/UNEP analysts studied the manner in which agroclimatic zones within Kenya, each suitable for different crops, respond to years with high and low rainfall. Two examples are shown in Figure 3.4. There is a marked contrast between the years of abundant and limited water supply. If conditions typical of 1984 were to become the norm, much of the region would be forced to shift to livestock production and ranching, forsaking maize, cotton, coffee and tea.

Could farming practices change in time to avoid substantial suffering? Modern variants of traditional coping strategies may be of value in the short-term but, in many areas, their effectiveness has been reduced by social change and environmental degradation. In the longer term, adaptation will

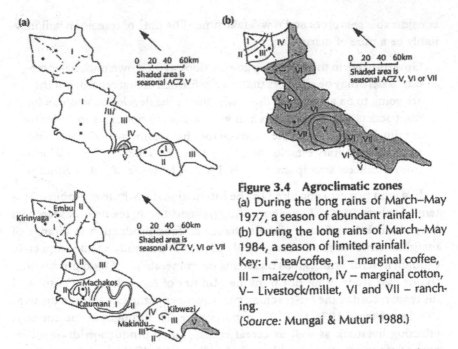

Figure 3.4 Agroclimatic zones
(a) During the long rains of March–May 1977, a season of abundant rainfall.
(b) During the long rains of March–May 1984, a season of limited rainfall.
Key: I – tea/coffee, II – marginal coffee, III – maize/cotton, IV – marginal cotton, V– Livestock/millet, VI and VII – ranching.
(*Source:* Mungai & Muturi 1988.)

depend on the availability of resources. In the richer, industrialized nations, it is quite likely that financial and technical resources will be available, although even here time will be of the essence. It takes ten years or so to develop a new variety of crop, even with the new biotechnologies (see Ch. 5), and about the same amount of time to reorientate a farming system from production through marketing to the consumer. In the Third World, such resources are unlikely to be available to all but the rich.

With harvest failure an increasing occurrence, will Third World nations be able to rely on surplus production elsewhere being available for purchase or as aid? The lessons of the early 1970s suggest that this would be unlikely to be the case (Kelly et al. 1983). Extreme weather and climate affected many parts of the world during that period, and pressure on world reserves of grain mounted. To meet the 12 per cent shortfall in their harvest, the Soviet Union bought almost 30 million tons of grain from the USA and Canada in late 1972. The purchase was made on the private market and was shrouded in secrecy. The result was panic, hoarding and speculation. World food reserves crashed to three weeks' supply, and prices eventually rose to four times their previous level. Poorer nations had no hope of buying food to supplement local supplies. The richer nations were loath to release dwindling food reserves as aid.

The North American grain belt provides much of the surplus production that supports world reserves of grain. If drought becomes a more frequent occurrence in that part of the world, the same process may occur again. In the aftermath of the drought of 1988–9, the amount of wheat available from the USA under donation and concessional aid programmes was cut by 10 per cent, as supplies fell to their lowest level since the mid-1970s. Buying food on the market to meet this shortfall was not considered an option, as it would have boosted prices, further aggravating the problem. The economic forces that shape the availability of food are extremely sensitive, amplifying the impact of climatic change and driving prices beyond the reach of all but the wealthiest nations. The loss of world food reserves could well prove the most immediate and wide-ranging consequence of global warming (Brown et al. 1989).

Conflict and refugees

The impact of climatic change and sea-level rise will cascade through the social and economic structures of the nations of the world. The result will be profound disruption to Third World economies already struggling to achieve growth. "Few outside the industrial world would have the structure or resources to manage a continuing crisis. Disorder, terrorism, civil war, economic breakdown or even bankruptcy could become endemic" warns Sir Crispin Tickell, Britain's former ambassador to the United Nations. It may be that economic warfare breaks out as old patterns of trade die and new ones emerge, as nations attempt to stockpile remaining resources and others exert pressure to make them available.

At the community level, environmental pressure will combine with social injustice to increase competition for resources. Social unrest will be the inevitable result. The link between economic development and the state of the environment is now widely accepted, yet few have realized that global warming also underlines the intrinsic link between environmental degradation and the battle for human rights. "At the root of this environmental problem is a land problem that has to be solved if any serious ecological policy is to be taken", according to Julio M. G. Gaiger, President of the National Indian Support Association in Brasil. For those who despair of the struggle, migration may be the only choice. According to Jodi Jacobson of the Worldwatch Institute, "a one-meter rise in ocean levels worldwide . . . may result in the creation of 50 million environmental refugees from various countries – more than triple the number in all recognized refugee categories today" (Jacobson 1989). Migration is already a common response to stress and disaster, but it is usually confined within national boundaries. The geographical scale of the impact of global warming means that transborder migration could become commonplace – from Bangladesh to India, from North Africa to southern Europe, from Central America to the USA. Present-day refugees tax the resources of host countries. How will the flood of environmental refugees be met – with compassion or resentment?

Conflict between nations is a real possibility. Mostafa Tolba, former Executive Director of the United Nations Environment Programme, gives a graphic example:

> I come from Egypt and Egypt is in a dry part of the world. The Nile gets its water resources from the uphills in Africa . . . If some country upstream puts a dam on the Blue Nile that brings almost 80 per cent of the water of the Nile down to Egypt, do you believe that 52 million people are going to sit back and wait, seeing themselves dying because of no water?

We allow this to happen today . . . .

For many in the Third World, life is already a struggle for survival. It is against this backcloth that the impact of global warming must be judged. Global warming will add further stress to a situation unbearable to many even in the present-day. "In terms of absolute numbers there are more hungry people in the world than ever before, and their numbers are increasing . . . The gap between rich and poor nations is widening – not shrinking – and there is little prospect, given present trends and institutional arrangements, that this process will be reversed", comments the World Commission on Environment and Development (WCED 1987). Global warming provides a graphic illustration of the direction in which current trends are leading us, of a future in which the famines of northern Africa, the floods in Bangladesh, the storm damage in the Caribbean, may become an everyday occurrence in one part of the world or another.

Although there may be uncertainties concerning the detailed impact of global warming, it is possible to forecast with absolute precision those who will experience the greatest impact of global warming. It will be those already living on the margins of survival as a result of environmental stress, economic disadvantage or social injustice. Some people, some nations, may benefit from the change in climate; their climate may become more benign. They will have the resources to adapt. However, those who are poor, those who are vulnerable, will inevitably suffer, finding their prospects for survival becoming slimmer and slimmer. This will be the most important impact of global warming. It will not be measured in terms of degrees Celsius or millimetres of rainfall; it will be measured in terms of human life.

In their actions to achieve the objective of the Convention and to implement its provisions, the Parties shall be guided, *inter alia*, by the following:

- The Parties should protect the climate system for the benefit of present and future generations of humankind, on the basis of equity and in accordance with their common but differentiated responsibilities and respective capabilities. Accordingly, the developed country Parties should take the lead in combating climate change and the adverse effects thereof.
- The specific needs and special circumstances of developing country Parties, especially those that are particularly vulnerable to the adverse effects of climate change, and of those Parties, especially developing country Parties, that would have to bear a disproportionate or abnormal burden under the Convention, should be given full consideration.
- The Parties should take precautionary measures to anticipate, prevent or minimize the causes of climate change and mitigate its adverse effects. Where there are threats of serious or irreversible damage, lack of full scientific certainty should not be used as a reason for postponing such measures, taking into account that policies and measures to deal with climate change should be cost-effective so as to ensure global benefits at the lowest possible cost. To achieve this, such policies and measures should take into account different socio-economic contexts, be comprehensive, cover all relevant sources, sinks and reservoirs of greenhouse gases and adaptation, and comprise all economic sectors. Efforts to address climate change may be carried out co-operatively by interested Parties.
- The Parties have a right to, and should, promote sustainable development. Policies and measures to protect the climate system against human-induced change should be appropriate for the specific conditions of each Party and should be integrated with national development programmes, taking into account that economic development is essential for adopting measures to address climate change.
- The Parties should co-operate to promote a supportive and open international economic system that would lead to sustainable economic growth and development in all Parties, particularly developing country Parties, thus enabling them better to address the problems of climate change. Measures taken to combat climate change, including unilateral ones, should not constitute a means of arbitrary or unjustifiable discrimination or a disguised restriction on international trade.

Article 3, Principles, Framework Convention on Climate Change

The framework convention on climatic change

The Framework Convention on Climate Change will shape the global response to climate change over years to come. The stated objective of the Convention is the "stabilisation of greenhouse gas concentrations . . . at a level that would prevent dangerous anthropogenic interference with the climate system". This should be achieved "within a time frame sufficient to allow ecosystems to adapt naturally . . ., to ensure that food production is not threatened and to enable economic development to proceed in a sustainable manner". The interpretation and implementation of the measures embodied in the Convention are to be guided by a set of principles, included at the insistence of the G77 Group of developing nations. To have guiding principles stated in this fashion is unusual, if not without precedent, in a convention of this nature. Given the ambiguity in various sections of the Convention, these principles may have an important role to play. The Convention does not spell out specific emission control targets. This is to be left to the protocols that will accompany the Convention. However, it does state that stabilization of emissions at 1990 levels by the year 2000 is desirable, with the burden of responsibility falling to the industrialized nations. This latter requirement is the main manifestation of the North–South divide that underlies the problem of global warming.

In this section, we discuss the Framework Convention, drawing out various aspects of the climate debate that have particular relevance for Southern nations. We pay particular attention to the issues of precautionary action, responsibility, the specific situation of the Third World and climate aid. See relevant sections of Grubb (1991) and Mintzer (1992) for a summary of these and other socio-economic and political issues raised by global warming. Finally, we list the commitments that developing nations take on when ratifying the Convention.

The need for precautionary action

The issue of uncertainty – and the amount of time it will take to reduce the uncertainties significantly – is a critical characteristic of the global warming issue. It has underlain much of the debate concerning the need to respond to the threat. It is impossible to forecast with precision the societal trends that will determine the amount of pollution in the atmosphere. The computer models used to forecast the effect on climate – based on the models used to predict the weather from day to day – are in the early stages of development, and understanding of the factors that will decide the ultimate impact on regional climate is poor. Decision-makers have to resolve just what can and should be done in view of these uncertainties.

It has been argued that the science of global warming is too uncertain to warrant action at this time. This attitude had a strong influence on the Bush Administration in the USA during the negotiating of the Framework Convention and has, in some quarters, been interpreted as a case for dismissing global warming entirely as a cause for concern. "The notion that 'future risk cannot be estimated with certainty' is sometimes translated, by a curious leap of faith, into the assertion that 'there is no future risk' ", observed Richard Benedick, former US Deputy Assistant Secretary for the Environment. The experts who compiled the Commonwealth Secretariat report on global warming firmly rejected "the policy of doing nothing". Noting that "such an approach has been followed all too often by governments in the past", they argued that "delay now can only magnify the scale of the ultimate damage and emergency and the attendant costs" (Anon 1989).

There is, however, a more substantive point raised by the issue of uncertainty. Responding to global warming by reducing greenhouse gas emissions or planned adaptation could be a costly affair. How far should we go along this path, given that the projections do contain many uncertainties? The response adopted by the international community is that precautionary action is warranted at this time. Having accepted that the potential impact of climate change is considerable, the risks associated with inaction are too great to countenance. On the other hand, it is difficult to justify substantial expenditure when the threat is poorly defined. Measures should, therefore, be carried out that are zero- or low-cost, or else have compensatory benefits in other areas, reducing other environmental problems or supporting the process of sustainable development. As scientific understanding develops, it is possible that the threat may prove to be not as serious as currently envisaged. In which case, the cost of responding unnecessarily will have been minimal. If the threat of climate change does, indeed, prove serious, then it will be that much easier to take further action, as a start will have been made.

Of course, translating the precautionary principle into action raises many other questions concerning the identification of appropriate measures and the estimation of costs. However, it is striking that many of the measures that might be taken in response to global warming do make good sense anyway, amounting to the wise and sustainable management of finite resources.

Responsibility for global warming

Participants at the New Delhi Conference observed, with grim irony, that the effects of global warming will be greatest in the South, whereas the source of the problem is the pollution caused by the industrialized North. As the authors of a recent report by the US Environmental Protection Agency observe, "most of the greenhouse gas emissions currently committing the world to climate change can be traced to activities by the industrialized countries" (Lashof & Tirpak 1990).

The production of energy from fossil fuels (coal, gas and oil) is the major single source of greenhouse gases. In the early 1990s, energy generation released over 6000 million tonnes of carbon into the atmosphere, as well as other greenhouse gases such as methane and nitrous oxide and the pollutants that generate ozone. Per head of population, emissions from the industrialized nations far outweigh the contribution of the Third World. A citizen of the USA typically generates five tonnes of carbon per year, compared with half a tonne produced by a Chinese citizen and a third of a tonne by a Brasilian (Marland 1989). Over the period since 1870, the developing world has been responsible for no more than 15 per cent of total carbon dioxide emissions (Pachauri 1989). The industrialized nations, with about a quarter of the world's population, are directly responsible for well over a half the greenhouse gas emissions causing global warming (Fig. 3.5). The USA –

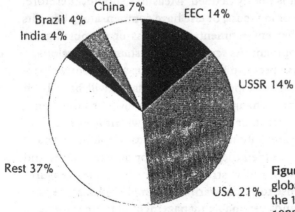

Figure 3.5 Contribution to global warming by region during the 1980s (source: Lashof & Tirpak 1989).

with a mere 4 per cent of the world's population – contributes over a fifth of the global total. It could also be argued that the industrialized world is indirectly responsible for much of the emission nominally attributed to the Third World, through the activities of multinational corporations, certain forms of development assistance, economic policy and unfair trade practice.

In what has been considered an attempt to deflect attention away from the indisputable fact that global warming is the responsibility of the industrialized world, considerable emphasis has been given to the future contribution of the Third World – which will rise considerably if, in the process of industrialization, it follows the past example of the North. "Developing countries increasingly find themselves put in a position where they are made responsible for global environmental problems they did little to create," observes Sir Shridath Ramphal, former Secretary-General of the Commonwealth. A significant rise in the contribution of the Third World is predicted, based on the assumptions of rapid population growth, the adoption of Western models of industrialization and continued economic development. There can be no doubt that population growth rates will be highest in the Third World, but against this increase in population must be weighed the gross inequality in emissions per head of population. The immediate problem is not the potential contribution of the Third World but the conspicuous overconsumption of the North in the present day. It is highly unlikely that the Third World will come anywhere close to rivalling Northern emission rates per head of population in the foreseeable future. Consumption- or GDP-weighted schemes have been proposed for the equitable allocation of emission reduction targets.

> While the historical contribution of the developing world to the problem is trivial compared to that of the industrialized nations, this situation may change as rising populations and the desire to improve living standards results in a substantial increase in the Southern contribution over the coming decades. It is essential that the developing world be involved in international measures to control emissions, adopting sustainable methods of development, which do not repeat the mistakes made by the North.   Suzanne Mubarak, First Lady of Egypt

The Third World must be supported in its efforts to limit its growing contribution to global warming through the adoption of sustainable forms of development. However, the weight of responsibility lies with the North. "The cost of limiting emissions of greenhouse gases has to be borne by those societies who are responsible for the cumulative increase in their concentration levels", concludes Rajendra K. Pachauri, Director of the Tata Energy Research Institute in New Delhi.

On the basis that the "polluter must pay", it is only right and proper that the North should bear the costs of its past pollution of the atmosphere and take the initiative in controlling greenhouse gas emissions. Although the Framework Convention does not explicitly acknowledge the "polluter must pay" principle, it does cite the particular responsibility of the industrialized nations and the need for burden-sharing according to capabilities. Concern over the future role of the developing world is also recognized. In a curiously worded paragraph, the Convention makes the point that stabilization of greenhouse gas emissions will show "that developed nations are taking the lead in modifying longer-term trends . . . consistent with the objective of the Convention." This is implicit recognition of the power of the position many Third World nations have taken on climate change: limited involvement unless those developed nations with historical responsibility for the problem take action first. Is the primary motivation for emission control on the part of the industrialized world concern for the planetary environment or a less than altruistic fear that rising emissions in developing nations may exacerbate the problem in decades to come? The irony is that, given the state of development of the energy sector, stabilization of carbon emissions by the major industrialized nations is not too different from what is likely to happen anyway, even if no concerted action is taken to control emissions. Stabilization may not be a particularly convincing achievement, representing little more than "business as usual."

> The greatest environmental damage is occurring in developing countries. These countries are poor. They are faced with falling commodity prices, rise in protectionism, crushing debt burden and dwindling or even reverse financial flows. If the commodities bring little money, they must produce more of them to bring in the same amount and at times even less. To do this, they cut down trees, bring under cultivation marginal land, overgraze their pastures and in the process make desert out of previously productive land. But in these actions the poor have no choice. They cannot exercise the option to die today so as to live well tomorrow. Developing countries are caught in a vicious circle . . . To ask us to plan for our survival tomorrow when our survival today is in doubt is to demand too much of us. For it is only when we can survive today that we can talk of a tomorrow.
>
> Robert Mugabe, Prime Minister of the Republic of Zimbabwe

Realizing that it is likely to be more cost-effective to reduce emissions in the Third World than in the industrialized nations, because of the relative development of the respective energy sectors, it has been proposed that Northern nations can gain credit for emission reductions by financing control measures in the developing world. The process is known as joint implementation and it has raised critical questions regarding responsibility and equity. According to Grace Akumu of Climate Network Africa:

developed country Parties [to the Convention] should take the lead in combating climate change since they bear the greatest responsibility for the problem . . . If joint implementation were to benefit all countries equally, they would all have to start from equal negotiating positions. By treating all countries in a blanket manner under the principle of cost-effectiveness, joint implementation becomes another vehicle for perpetuating inequity.

It is argued that joint implementation will leave the North to continue its "profligate" lifestyle – the main cause of the global warming problem – and that the South is not in a strong enough position to ensure that joint implementation deals work in its favour.

The specific situation of the developing world

Global warming presents the Third World with both a challenge and an opportunity. The challenge arises because the impact of climate change and sea-level rise is likely to be most severe in the developing countries, already stressed by social and environmental problems. Moreover, the requirements of the Framework Convention may mean significant, and perhaps costly, changes in development planning to control releases of the major greenhouse gases. On the other hand, global warming provides the Third World with a lever that could be used to extract concessions from the industrialized world in terms of technology transfer, additional aid, trade agreements and debt relief. More generally, it highlights the inequality that currently exists between North and South, placing this issue firmly on the political agenda.

Intense lobbying by Southern delegates ensured that the specific situation of the Third World was embodied in the Framework Convention. Reference is made to:

- the need for additional financial resources and technology transfer to assist the developing nations meet their obligations under the Convention
- the fact that economic and social development and poverty eradication are the first and overriding priorities of the developing nations
- the specific needs of particular developing nations such as small island States and those with low-lying coastal areas, areas prone to natural disasters, liable to drought and desertification, and so on; and
- the situation of countries vulnerable to the adverse effects of measures taken in response to global warming (e.g. fossil fuel controls).

In specifying commitments under the Convention, a distinction is drawn between the industrialized nations and the developing world (with some account taken of the situation of the "economies in transition", the former centrally planned nations).

> It is . . . of utmost importance that development should be viewed in the widest possible context. The best way to appreciate this is to accept the concept of limits, that the planet's resources are finite, that the technological tools to exploit these resources are also limited at any particular time and that the distribution of both the technology and the resources are uneven and skewed as between developed and developing countries. Until recently, man has looked at the planet as large and with infinite resources that would not be affected by his activities. However, we have seen during the past two decades how the world has gone from one development crisis to another, and quite a few of these crises have been due, in no small measure, to a disregard for the need to incorporate environmental dimensions in the development process.
>
> Adebayo Adedeji, United Nations Under-Secretary-General

Nevertheless, as we shall see later in this section, the commitments that developing nations take up in ratifying the Convention are not trivial. Consequently, the issue of financial support has been a crucial one during the climate treaty negotiations.

The position of many Third World negotiators has been that they will not commit themselves to measures to limit global warming unless the industrialized world covers the additional costs of these actions. In principle at least, this has been accepted and a financial mechanism has been established to cover the additional costs of appropriate measures. This provision is in recognition of the fact that the North has the resources to combat climate change and that, if it does not assist the developing nations, their actions will exacerbate the problem; it is not acceptance of the "polluter must pay" principle.

For the time being, the climate fund will be administered by the Global Environment Facility (GEF). The aim of the Facility is to provide a channel of action through which the international community can assist developing nations to deal with global environmental problems. Officially established in November 1990, the GEF was developed by representatives of both industrialized and developing countries with the World Bank, the United Nations Development Programme and the United Nations Environment Programme. There are four areas of specific concern:

- reducing and limiting emissions of greenhouse gases
- preserving the Earth's biological diversity and maintaining natural habitats
- arresting the pollution of international waters, and

- protecting the ozone layer from further depletion.

To qualify for a grant or low-interest loan, the per capita gross domestic product of the applicant country must have been at or below US$4000 in 1989. It must also be shown that the proposed project would not be economically viable for that country without GEF support.

Whether the climate fund remains under the GEF umbrella remains to be seen as there is considerable suspicion of the Facility among Southern nations, not least because of the control exerted by the World Bank. It is debatable whether the GEF can be considered to have an "equitable and balanced representation of all Parties within a transparent system of governance", in the words of the Framework Convention. There is also some doubt concerning the extent to which effective support can be given through a multilateral fund such as the GEF given the track-record of large-scale development assistance (McCully 1991).

Developing nation commitments

Although the Framework Convention places the emphasis on the industrialized world in specifying measures to be taken in response to global warming, developing nations do take on a fairly substantial list of obligations. They are obliged to develop and publish national plans that contain measures to mitigate climate change by addressing emissions of greenhouse gases and their removal from the atmosphere by sinks. National inventories of emissions by sources and removal of greenhouse gases by sinks are required to achieve this end. In common with the other Parties, they must:

- co-operate in the development, application and transfer of technologies that control, reduce or prevent the emission of greenhouse gases
- promote the sustainable management of sinks and reservoirs (forests, oceans, and so on)
- co-operate in preparing their adaptation to the impact of climate change (including integrated plans for coast and air management, water resources and agriculture) and the protection and rehabilitation of areas affected by drought and desertification
- undertake environmental impact assessment of policies and projects that may affect climate change
- promote and co-operate in scientific, technical and other research, systematic observation (or any other method to reduce or eliminate remaining uncertainties regarding the causes, effects, magnitude and timing of climate change)

- promote and co-operate in the full and open exchange of relevant information
- promote and co-operate in education, training and public awareness, and
- communicate to the Conference of the Parties information relating to implementation.

As James Cameron has observed, "there is a lot here for any government to do, whether developed or developing, and I imagine that many administrations around the world will find it extremely hard to comply with these obligations. At the very least, they will need a great deal of assistance and advice" (Cameron 1993). Various initiatives have been established by relevant United Nations agencies and by other bodies to support this process. It remains to be seen whether they will be sufficient to ensure an effective Southern response.

Responding to global warming

As principle becomes translated into action, difficult decisions will have to be taken regarding the most appropriate means of responding to global warming. Given the wide-ranging nature of the source of global warming and of its impact, a broad strategy covering both emission control and adaptation is needed. Moreover, the fundamental link between global warming and the process of economic growth means that any strategy will have substantial implications for the development process.

Flavin (1989), and relevant sections of Leggett (1990) and Mintzer (1992), provide a comprehensive introduction to emission control and related socio-economic and political issues. For a more technical introduction, see Lashof & Tirpak (1990). The general conclusions of the IPCC assessment, which have guided the formulation of the Framework Convention on Climate Change, are presented in the 1991 IPCC report (Intergovernmental Panel on Climatic Change 1991). For a general discussion of the issue of adaptation, see Karas & Kelly (1989), National Academy of Sciences (1991) and Turner et al. (1990). The conclusions of the IPCC assessment concerning adaptation are contained in Tegart et al. (1990) and the IPCC Response Strategies Report (Intergovernmental Panel on Climatic Change 1991). For a developing country perspective, see Jodha (1989).

In all likelihood, emission control will not prove effective in limiting the impact of global warming over the next few decades, as the reductions needed to slow the rate of warming are considerable. Identifying and implementing measures to ease the process of adaptation to the changing environment must be a priority. All too often, adaptation is a traumatic process triggered by disaster. In some areas, where planning timescales are relatively short, environmental change may be accommodated as it occurs. In other areas – such as the protection of natural resources, irrigation, agricultural land use, coastal engineering and energy planning – timescales are of the order of decades, and anticipation of change will be crucial. Insofar as it is possible, plans for long-term projects must be "climate-proofed." A comprehensive environmental impact assessment of any development project must consider not only the effect of the project on the environment but also the implication of changing climate and sea level for the effectiveness of that project.

Planning for change is rendered difficult by the many uncertainties that affect estimates of the impact of global warming. Lacking detailed predictions, maintaining flexibility will increase options in the face of climate change and sea-level rise. The Commonwealth Secretariat experts recommend that:

> to maintain maximum flexibility, priority should be given . . . to enlarging long-term options for adjustment to changing climate, and gathering and conserving genetic materials and knowledge of plants, animals and agricultural practices that have been valuable under conditions of uncertainty and climatic variability. (Commonwealth Group of Experts 1989). In addition, improving resilience (the degree to which natural and human systems can cope with environmental stress) will reduce impacts, whatever their nature may be. Inevitability, the suitability of different adaptive strategies will vary from region to region, depending on the nature of the impact, resource availability, and the prevailing social, economic and cultural conditions. This means that a local approach is likely to be the most effective, implying the active awareness of involvement of local scientists, engineers, planners and communities.

A variety of adaptive responses have been recommended (see Table 3.1 for examples). Some are likely to prove costly, whereas others can be achieved with a very modest investment. In most cases, measures that ensure flexibility and resilience would be of immediate benefit, reducing vulnerability to present-day social, economic or environmental threats, whatever the ultimate impact of global warming. They represent a sensible first step in a precautionary response to adapting to climate change. In essence, they amount to better management of the Earth's limited resources, consistent with the principles of sustainable development.

Table 3.1 Areas in which adaptive responses to global warming should be considered.

Improve water management	Increase options	Improve food distribution
– Irrigation provision and scheduling	– Environmental monitoring	– Transportation systems
– Windbreaks	– Application of climate information	– International trade
– Snow management	– Use of appropriate technology	– Price control
– Flood control	– Communications	
– Reservoirs	– Education and raising awareness	
Improve agricultural practice	Climate-independent food production	Improve coastal policies
– Soil management	– Environmental control	– Land-use zoning
– Crop rotation	– Aquatic management	– Coastal defence
– Intercropping	– Chemical foods	– Flood control
– Selective use of agrochemicals		– Flood warning systems
– Scheduling		
Exercise crop selection	Maintain food reserves	– Develop international mechanisms
– Crop changes to favour resilience	– Food stores	– Aid and disaster relief
– Genetic research	– Reduce storage losses	– Equitable trade
	– Livestock	– Technology transfer
	– Preserves	– Subsidies
		– Compensation and reparation
		– Debt relief

The measures needed to reduce greenhouse gas emissions are well established. Greater use of energy-conservation technology would reduce carbon emissions, as would improving energy efficiency, switching to zero- or low-carbon emitting fuel sources (such as the renewables) and slowing deforestation rates. Carbon could be removed from the atmosphere by reforestation or afforestation. Land management could be improved to protect carbon sinks. Methane emissions can be reduced by reducing releases from landfill sites, leakage from gas pipelines and curbing agricultural releases where it is possible. Reducing halocarbon use would protect the ozone layer as well as limit the threat of climate change.

Although precautionary action may be relatively easy to carry out, a large-scale cut in emissions, if deemed necessary, could prove extremely difficult. In the case of carbon dioxide, global emissions would have to be cut by over 50 per cent to limit the rise in atmospheric concentrations significantly. This would require a major shift away from fossil fuel combustion. Nevertheless, from a technical standpoint, significant reductions in emissions could be achieved. The difficulties, though, lie in the social, economic and political barriers to be surmounted. For this reason, this section focuses not on the technicalities of control and adaptation but on the underlying issues that will critically affect the global response to the challenge. Perhaps the most fundamental criticism of the Framework Convention on Climate Change is that it addresses the response to global warming in isolation, without full consideration of the mesh of social, economic and political issues within which an effective response to problems must be located.

We must remember that there is only one global atmosphere and it belongs to all nations of the world. Atmospheric pollution in California or Scotland will affect Kenya as well as other parts of the world. Every nation must therefore join in the worldwide effort to protect the atmosphere in order to maintain a habitable world.

Herick Othieno, Kenyatta University

It remains an article of faith that economic growth must be based on a significant increase in energy consumption. This is despite the fact that the historic link between energy use and gross national product (GNP) was broken in many industrialized nations during the late 1970s and early 1980s (Goldemberg et al. 1988). Over that period, GNP in the nations of the Organization for Economic Co-operation and Development rose by 21 per cent, yet total energy use declined by 6 per cent. That development was triggered by the oil crisis of the 1970s and it rested on greater use of energy-saving technology, and on the elimination of waste, overconsumption and inefficiency. With easy access to what were seen as virtually unlimited resources, the Northern nations have been profligate in their use of energy. It is now clear that industrialization powered by the inefficient consumption of fossil fuels cannot be considered a sustainable form of development. The political and economic costs became apparent in the 1970s. The environmental costs are now manifest in the threat posed by global warming. However, is there an alternative?

In their pioneering work, *Energy for development*, Jose Goldemberg and his colleagues showed that it would be technically and economically feasible for developing nations to raise their living standards to western European levels with an increase in energy consumption per head of population of only 30 per cent, well below conventional estimates (Goldemberg et al. 1987). This alternate perspective results from looking at energy needs not from the point of view of supply – the standard approach, which often leads to the construction of large, centralized, and often inefficient, power supply systems benefiting industry and the urban elite, rarely the poor – but from the point of view of the use to which energy is put. It means considering in detail where and how energy is used, rather than crudely estimating a nation's total energy needs and then constructing an expensive and inefficient centralized energy supply system to meet that projected demand. From this perspective, the value of energy-saving technology is clear. Money invested in increasing energy supply would, in many instances, be better spent on improved cooking stoves, more efficient light bulbs, biogas plants, producing gas generators and modernizing bioenergy. Goldemberg and his colleagues argue that their "end-use" approach has other advantages too. It would reduce dependence on imports of coal or oil, increase self-reliance, and make more effective use of human labour. It would ensure that the interests of those most in need are not neglected.

There is a marked contrast between this end-use approach and the "trickle-down" policies advocated by the International Monetary Fund (IMF). The approach of the IMF and many other agencies has been to target financial assistance on the industrial sector in the belief that benefits will then spread throughout the economy. With debts rising, capital in flight and commodity prices falling, all that has trickled down to the poor is the impact of austerity programmes and rampant inflation, leaving them with little option but to exploit what natural resources remain (George 1988).

If the Third World is to adopt a new form of development that is sustainable, ensuring environmental protection as well as economic growth, the nature of development aid will be crucial. Although both the World Bank and the IMF have expressed their new-found awareness of environmental concerns in recent years, there is still little evidence that this awareness is being reflected at the policy level. The Earth Summit in June 1992 called for a supplement to the International Development Association (IDA), the subsidized loan fund for the lowest income developing countries, to support environmental protection measures. World Bank proposals for the fund (the Earth Increment) cover renewable energy sources and clean coal technologies, but explicitly exclude energy efficiency, since low-income countries use "comparatively little energy for industry, cars, and household purposes". The fact that the best time to ensure high standards of energy conservation is during the construction of industry, transportation systems and so on appears to have escaped the Bank's analysts.

> Current development practice is based on a model that demeans the human spirit, divests people of their sense of community and control over their own lives, exacerbates social and economic inequity, and contributes to the destruction of the ecosystem on which all life depends.   The Manila Declaration on People's Participation and
> Sustainable Development, Manila, 10 June 1989

The international dimension is crucial. "Developing countries face the dilemma of having to use commodities as exports, in order to break foreign exchange constraints on growth, while also having to minimize damage to the environmental resource base supporting this growth", comments the World Commission on Environment and Development (WCED1987). Intrinsic to the dilemma is that commodity prices rarely reflect the true value of the resources depleted during their production. The Third World is expending its environment as a subsidy to the North. This problem has been further aggravated by the substantial fall in the prices of most major Third World commodities over the past ten years, reflecting the inequity inherent in any trade with the North.

The only solution, according to the World Commission on Environment and Development, is to develop an international system that fosters sustainable patterns of trade and finance. The World Commission has recommended that the mandate of multilateral trade forums such as GATT (the General Agreement on Tariffs and Trade) and UNCTAD (the United Nations Conference on Trade and Development) be altered to reflect the basic principles of sustainable development. It has also proposed that the code of conduct for transnational corporations, under discussion by the United Nations, be expanded to cover environment and development. Beyond this, a new form of accounting must be developed in which external costs such as environmental damage and resource depletion are accurately reflected in pricing rather than conveniently ignored (Pearce et al. 1989). The environment has for too long been considered a free resource by both North and South. Global warming highlights the unsustainable nature of this short-sighted attitude.

Whether or not international inequities can be reduced, additional financial resources are going to have to be found if the Third World is to limit pollution, protect its environment and adopt new sustainable forms of growth. Arguably the most direct way to achieve this end would be to halt the drain on Third World economies by eliminating the crippling burden of debt. In a sense, providing additional aid for climate control, or any other form of environmental protection, is to treat the symptom rather than the cause. The result of the reduction in public spending and the concentration on export earnings – the "adjustment" measures devised by the IMF to deal with the debt crisis – is to force severe restraints on support for environmental protection and to hasten environmental degradation. If that pressure is to be relieved, the debt crisis in the South must be resolved. Northern governments have been reluctant to confront the debt issue when considering responses to global warming. Yet participants at the International Conference on Global Warming and Climatic Change (African Perspectives), held in Nairobi in May 1990, ranked debt and fair trade the highest priority in their call for action on the part of the industrialized nations (Ominde & Juma 1991).

At the moment we have a common future only in extinction. In the immediate five to ten years people in agribusiness do not have a common future with the peasant in Asia or Africa who is dying. A peasant in India about to be displaced by a dam funded by the World Bank does not have a common future with the fellow in Washington with their thousands of dollars salaries. But the theme of a common future has been used to allow the dominant powers and the industrial interests they serve to say: "Listen, there's a common global crisis – we've got the funds, we've got the technology, we've got the answers." It is a way to appropriate the terms of ecological resistance and so to regain control.

Vandana Shiva, Research Foundation for Science and Ecology

A common future?

Action to limit the threat of climate change must be seen in the context of the broader principles of sustainable development – meeting the needs of the present without jeopardizing the interests of future generations (WCED 1987). And, if equity between generations is a goal, then equality within this generation must be an immediate priority. Indeed, if this issue is not addressed, there can be no effective response to global warming. The root causes of climate change are to be found in the inequity and misplaced priorities that characterize our modern society.

When you have the North fixing the prices of the primary commodities and they fix the prices of the manufactured goods that we buy, you can't win! How do you win? They fix the price at which they buy our commodities and they fix the price at which they sell their commodities to us. What kind of arrangement is that? Even if you dealt with an angel you would still run into trouble. But you don't deal with angels. You are dealing with people who say making a profit is a religion.

Julius Nyerere, former President of Tanzania

Whether the result of historical accident or exploitative design, the industrialized world consumes far more than its fair share of the Earth's resources (Trainer 1989). It is Northern overconsumption, waste and inefficiency that is driving global warming and the inequities between North and South that are forcing Southern nations to destroy their forests and sell off their natural resources at unreasonable prices. The causes of global warming can be found, too, in the misplaced priorities typical of modern society, in the emphasis on guns rather than butter, meat rather than grain. The world has been spending over US$2000 million a day on its military establishments, dwarfing expenditure on environmental protection, health care and humanitarian aid (Renner 1989). Even in terms of military goals, this diversion of resources makes no sense. What use a cruise missile if global security is threatened by climate? False perceptions of what constitutes security are diverting economic, natural, technical and human resources away from a far more pressing defence of our long-term interests.

In the long term, these are the issues that have to be addressed if global warming is eventually to be curbed. An equitable strategy for dealing with global warming must be based on the true principles of sustainable development. It should preserve ecological integrity, respect basic human rights, acknowledge cultural diversity and be founded in consultation, co-operation and accountability. If not, it is doomed to failure. The global nature of the threat posed by the greenhouse effect contains an imperative for international co-operation that is without precedent. That the threat of global warming encompasses the many environmental and social problems faced today – atmospheric pollution, desertification, malnutrition and starvation, economic disadvantage, injustice and inequity – provides a common focus for a concerted effort to improve the quality of life for all.

References

Anon 1989. *Climatic change: meeting the challenge*. Report by a Commonwealth Group of Experts. London: Commonwealth Secretariat.

Brown, L. R., C. Flavin, S. Postel 1989. A world at risk. In *State of the world 1989*, Brown, L. R. (ed.), 3–20. New York: Norton.

Cameron, J. 1993. Implementation of the Climate Convention. *Tiempo* 7, January.

Carter, T. R., M. L. Parry, S. Nishioka, H. Harawawa 1992. *Guidelines for assessing impacts of climatic change*. Oxford/Tsukuba: Environmental Change Unit/Center for Global Environmental Research.

Flavin, C. 1989. *Slowing global warming: a worldwide strategy*. Worldwatch Paper 91, Worldwatch Institute, Washington.

Foley, G. 1991. *Global warming: who is taking the heat?*. London: Panos Institute.

George, S. 1988. *A fate worse than debt*. London: Penguin.

Goldemberg, J., T. B. Johansson, A. K. N. Reddy, R. H. Williams 1988. *Energy for a sustainable world*. New Delhi: John Wiley.

Goldemberg, J., T. B. Johansson, A. K. N. Reddy, R. H. Williams 1987. *Energy for development*. Washington DC: World Resources Institute.

Gribbin, J. & M. Kelly 1989. *Winds of change*. Sevenoaks: Headway.

Gribbin, J. 1990. *Hothouse Earth*. London: Bantam Press.

Grubb, M. 1991. *Energy policy and the greenhouse effect*, vol. 1: *policy appraisal*. London: Royal Institute of International Affairs.

Harwell, M. A. & T. C. Hutchinson 1985. *Environmental consequences of nuclear war,* vol. 2: *ecological and agricultural effects* [SCOPE 28]. Chichester, England: John Wiley.

Houghton, J. T., G. J. Jenkins, J. J. Ephraums (eds) 1990. *Climate change: the IPCC scientific assessment.* Cambridge: Cambridge University Press.

Houghton, J. T., B. A. Callander, S. K. Varney (eds). 1992. *Climate change 1992: the supplementary report to the IPCC scientific assessment.* Cambridge: Cambridge University Press.

Intergovernmental Panel on Climate Change (IPCC) 1991. *Climate change: the IPCC response strategies.* Washington DC: Island Press.

Jager, J. & H. L. Ferguson (eds) 1991. *Climate change: science, impacts and policy.* Cambridge: Cambridge University Press.

Jacobson, J. L. 1989. Abandoning homelands. In *State of the world 1989*, L. R. Brown (ed.), 59–76. New York: Norton.

Jodha, N. S. 1989. Potential strategies for adapting to greenhouse warming: perspectives from the developing world. In *Greenhouse warming: abatement and adaptation*, N. J. Rosenberg, W. E. Easterling III, P. R. Crosson, J. Darmstadter (eds), 147–58. Washington DC: Resources for the Future.

Karas, J. H. W. & P. M. Kelly 1989. *The heat trap.* London: Friends of the Earth.

Kelly, P. M., D. A. Campbell, P. P. Micklin, J. R. Tarrant 1983. Large-scale water transfers in the USSR. *Geojournal* 7, 201–14.

Lashof, D. A. & D. A. Tirpak (eds) 1990. *Policy options for stabilizing global climate.* Washington DC: Environmental Protection Agency.

Leggett, J. (ed.) 1990. *Global warming: the Greenpeace Report.* Oxford: Oxford University Press.

Marland, G. 1989. *Fossil fuel CO_2 emissions: three countries account for 50% in 1986.* Oak Ridge: Carbon Dioxide Information Center, Oak Ridge National Laboratory.

McCully, P. 1991. The case against climate aid. *Ecologist* 21, 244–51.

Miliman, J. D. et al. 1988. *Environmental and economic impact of rising sea level and subsiding deltas: the Nile and Bengal examples.* Woods Hole: Woods Hole Oceanographic Institute.

Mintzer, I. (ed.) 1992. *Confronting climate change.* Cambridge: Cambridge University Press.

National Academy of Sciences (NAS). *Policy implications of greenhouse warming: report of the Adaptation Panel.* Washington DC: National Academy Press.

Ominde, S. H. & C. Juma (eds). *A change in the weather: African perspectives on climatic change.* Nairobi: ACTS Press.

Pachauri, R. K. 1989. Energy efficiency in developing countries: policy options and the poverty dilemma. Paper presented at the RIIA/BIEE/IAEE Fourth International Energy Conference on "Environmental Challenges: the Energy Response", London, December 4–5 1989.

Parry, M. L., T. R. Carter, N. T. Konijn (eds) 1988. *The impact of climatic variations on agriculture.* Vol. 1: *Assessments in cool temperate and cold regions*; vol. 2: *Assessments in semi-arid regions.* Dordrecht: Kluwer.

Pearce, D. W., A. Markandya, E. B. Barbier 1989. *Blueprint for a green economy.* London: Earthscan.

Renner, M. 1989. Enhancing global security. In *State of the world 1989*, L. Brown (ed.), 132–53. New York: Norton.

REFERENCES

Schneider, S. H. 1989. *Global warming*. San Francisco: Sierra Club.

Tegart, W. J. McG., G. W. Sheldon, D. C. Griffiths (eds) 1990. *Climate change: the IPCC impact assessment*. Canberra: Commonwealth of Australia.

Anon. 1992. International insurance pool proposal. *Tiempo* (4), 11–13.

Trainer, T. 1989. *Developed to death*. London: Merlin.

Turner, R. K., P. M. Kelly, R. Kay 1990. *Cities at risk*. London: BNA International.

WCED (World Commission on Environment and Development) 1987. *Our common future*. Oxford: Oxford University Press.

Wigley, T. M. L. & S. C. B. Raper. Implications of revised IPCC emissions scenarios. *Nature* 357, 293–300.

CHAPTER FOUR

Environmental reproduction and gender in the Third World

Cecile Jackson

People cannot change the way they use resources without changing their relations with each other.   Stretton (1976: 3)

Editors' Introduction

The Rio Declaration on Sustainable Development proclaims a goal of establishing a "new and equitable partnership" among States by means of new levels of cooperation and exchange. If that is to happen at the level of nations,
it must also occur within nations, between social groups and down to the
household and family level. Partnership, equity and discourse have featured
already in this volume: Piers Blaikie (Ch. 1) highlights the need for different
modes of analysis that promote discourse between the natural and social
sciences, and he describes a political-ecology model of people–environment
interactions; in addressing the central issue of sustainability, David Gibbon,
Alex Lake and Michael Stocking (Ch. 2) examine new paradigms of agricultural development that promote partnership, networks and systems-thinking
as means to towards a more just society and a friendlier use of the environment. In this chapter, Cecile Jackson emphasizes the gendered nature of
human interaction with the environment. She provides a gender lens through
which to see the population–environment debate, and a gender analysis
framework with which to study the differences and draw conclusions that are
policy-relevant.

Agenda 21 emphasized the establishment of sustainable livelihoods for the
up to 2000 million people who are impoverished and who are forced to
exploit their environments in ways that cannot possibly continue. Women
make up more than half these people. In many societies, they are the immediate exploiters of fuelwood, water and soil resources. Yet, in those same
societies, they are the marginalized and dispossessed. They are at once both

109

a cause of environmental degradation and a contribution to sustainable development. In its formulation, *Agenda 21* took up 27 principles of sustainable development, of which three are of direct relevance to gender differentiation and to women:

- through a recognition of the central role of women in development;
- through the need for citizen's participation;
- through an appreciation that human beings are at the centre of sustainable development.

The Rio Declaration argued for capacity building and the promotion of both ecologically and culturally sensitive schemes of agricultural and rural development. Included in capacity building is the role of local scientific and technical knowledge, in which women play a leading function. Cultural sensitivity means the understanding and incorporation of different perspectives, goals and interests, again an area where women play a leading role in sensitization as well as themselves being a key client group. Cecile Jackson shows in this chapter how these principles of sustainable development can be accommodated in a gender-analytical structure that shows the conflicts and struggles, as well as the shared and mutual interests of men and women. It is a vital and dynamic way of understanding environmental issues and enabling better choices to be made in development processes.

Introduction

This chapter attempts to show that environmental degradation embodies gender relations, and that a better understanding comes from viewing environmental change through the lens of gender concepts. In recent years there have been several analyses that have displayed the significance of equity issues in processes of environmental management and degradation. These have shown that the poor are most negatively affected by environmental degradation (Jodha 1986), that the poor are often blamed for resource degradation (for example Drinkwater 1989a), that resource depletion is often related to processes of accumulation, social differentiation and the spread of commodity relations (e.g. Blaikie & Brookfield 1987, Cliffe & Moorsom 1979) and that poverty drives rural people to exploit the environment (Blaikie 1986). However, class analyses of the causes and consequences of environmental degradation consider only one aspect of equity. People have gender identities too, as men and women, which constitute another axis of differentiation of equal importance in understanding changing environmental relations. A growing literature has documented the differential impact of environmental degradation on men and women, and several analyses have shown the differential participation by men and women in conservation projects and programmes. For example, Figure 4.1 shows the experience of one development project in which women's adoption of conservation technologies was generally (but not always) lower than men's. These kinds of differentials require explanation. A gender analysis offers frameworks for this, which can help to anticipate gendered constraints and opportunities in environment and development activities.

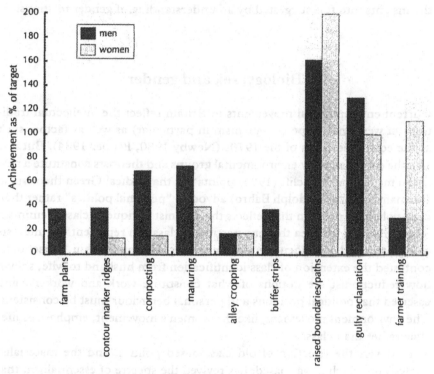

Figure 4.1 Adoption on soil conservation activities.

Gender analysis has emerged as part of a wider paradigm shift in the social sciences in which the relativity of peoples perceptions have become clear. For example, the environment is perceived very differently by rural women, elite urban women, rural men, forest officers, scientists and development agents. This shift has come about through both the realization that non-Western cultures are diverse and different from those of the West and through the crisis in science. In recent years many scientific certainties about environmental degradation (e.g. rates of soil erosion in Southern Africa, the primary status of West African forest and the concept of carrying

111

capacity in rangeland management) have been revealed to be questionable, not only by scientists with improved methods and data, but also by social scientists who have analyzed the sociology of science, the values it embodies, and the subjectivity of scientific methods and knowledge (Richards 1987).

This chapter will first set out the debates within feminism regarding the natural and the social, and it will then explore the concept of reproduction as developed in gender analyses of development. It will then examine the differential impact of environmental degradation on men and women, and the insights into this suggested by an understanding of gender relations.

Biology, sex and gender

Current environmental movements in Britain reflect the intellectual traditions of western Europe (Utopianism in particular) as well as factors such as the economic crises of the 1970s (Newby 1980, Pepper 1984). But it is also the case that many environmental groups and theorists constitute a new social movement. Redclift (1984) points out that radical Green theorists in Germany (such as Rudolph Bahro) advocate "personal politics" rather than class politics, a position that reflects the feminist critique of class. Feminism has dismantled the idea that any particular class can represent the interests of humanity. It has documented the power relations within households, contested the extension of class identification from husband to wife, shown how reductionist are notions of what constitute work and workers, and asserted that political positions and personal behaviour must be consistent. The environment movement, like the women's movement, emphasizes lifestyle rather than class.

However, the rejection of old class-based politics, and the materialist analysis on which it was based, has revived the spectre of essentialism; that is, the idea that individual identities reflect inherent characteristics or "essences". For example, ecofeminists suggest that women are inherently closer to nature than men (Shiva 1989). Thus, the debates about what is "natural" and what is "social" have been reopened (Agarwal 1991, Jackson 1992).

Environment, nature and conservation are seen from a sociological perspective not as givens but as ideas with their own history and politics. It is important to realize that the views we have of the environment are products of a range of social forces and are not based simply on "facts" that are true. In addition, this chapter holds that, although feminism has been influential in certain forms of environmentalism, there is a need to apply the ideas of gender analysis to our understanding of environmental degradation in the Third World to prevent the "politics of the personal" slipping into universalism and essentialism. This chapter will explain what such a gender analysis might involve.

In what way can our concept of nature be said to be a social construction? The idea of "nature" exists in Western thought in opposition to "culture" and the origins of this have been traced to the Scientific Revolution (Merchant 1980), further back to the impact of Christianity (White 1967) and even further back to transcendentalism in Greek traditions (Ruether 1979). Levi-Strauss (1969) suggests that the nature/culture dichotomy is a fundamental cognitive structure of thought in the human brain. Whatever the origins of the Western idea of nature, it certainly has a specific history and cannot be applied universally across all cultures and all times. Anthropologists have found a wide range of different understandings of, and means of bounding off, nature in other cultures (MacCormack & Strathern 1980). Within our own society, Nature has been seen as threatening and disorderly during the Scientific Revolution of the seventeenth century, exalted and mystical by the romanticist backlash, and a resource to be exploited in modern times. These views may coexist at any time, but shift in significance with tides of ideological and material change.

In addition, in Western history each age has expressed its understanding of nature anthropocentrically. For example, the concept of landscape and of conservation embodies human value-judgements about aesthetics and quality underneath their justification in scientific terms, such as the stability and diversity of climax vegetation. Similarly, the endangered species and animal rights movement embodies a profoundly anthropocentric perspective, and animals that humans find attractive, such as pandas, chimpanzees, seals and domesticated animals, are valued above insect life. Perceptions of nature not only reflect ideas about beauty and human empathy with humanoid forms of nature, they also embody class, race and gender identities. Thus, for example, the study of hunting in the British empire by MacKenzie (1988) shows the roots of modern African conservationism in the colonial hunting culture, with its upper-class, White, male, British role models and its fascination with Darwin and natural history. The perceptions of nature of White male settlers may have been specific to their social identities, but their economic dominance allowed boundaries between men and women, colonials and natives to be justified and reinforced by reference to these perceptions.

If historians put Western notions of nature in time contexts, anthropologists extend this analysis in contemporary societies of the South by questioning whether the nature/culture dichotomy is universal or uniform in meaning. Marilyn Strathern concludes that for Hageners (of Papua New

Guinea) "humanity is bounded off from the non-human, but does not seek to control it" (Strathern 1980: 218).

If we accept that nature is a social construct, what of sex and gender? The term gender is used to distinguish between, on the one hand, sex as the biological difference between men and women, and, on the other hand, the full identity as men or women formed and expressed through social processes. Thus, although the physical differences between the sexes are universal, gender is extremely variable between and within societies. Being a woman in one culture carries with it roles and expected behaviour very different from those of women in other cultures. Science has in the past assumed that women's biology constituted their destiny. Sayers (1982) details the intellectual contortions of respected nineteenth-century scientists who argued that women should not go to university since it would damage the development of their reproductive systems, and that women were less intelligent than men because of brain size. The latter argument moved through several phases. First, women were said to have smaller brains than men, but when it was realized that their brains were proportionately larger than men's, the goalpost shifted and the critical factor was said to be the size of the frontal area of the brain (larger in men). This argument held until it was found that Black people have large frontal areas of the brain. Clearly, racism and sexism were important dimensions of such "objective and value free" science.

Feminists have, in the main, thoroughly debunked the "natural" and the attempt to explain the subordination of women as biologically determined. Thus, the concept of gender asserts the social construction of male and female identities, and denies that the physical differences between men and women do not lead to "essential" masculine or feminine characteristics ordained by nature. One final comment on the use of the term gender. It is often remarked that gender is said to apply to both men and women, but in practice gender analysis focuses upon women. There is some truth in this. One reason is because of the need to redress the imbalance in research and information available about the two genders, but another reason lies in the manner in which humanity is characterized as male, and women exist in a sense in opposition to this. The male domination in the evolution of human cultures and their representations means that the masculine is the norm and the feminine is defined as different from this. The insistence by feminists that the use of the term "man" for both men and women obscures the way in which the term really does refer primarily to men is important. For example Barbara Rogers shows how the supposedly neutral measurement in terms of "man-hours" is not gender-neutral – a man working for one hour constitutes a man-hour, whereas in many calculations a woman working for one hour is counted as 0.6 of a man-hour (Rogers 1980). This is despite the absence of work studies to substantiate productivity differentials. The point can be further illustrated with reference to many definitions that imply male identity, such as "farmer" and "peasant", that require the addition of "wives" or "women" to neutralize the gender implications. If the content of gender analyses emphasizes women, this is in part because what is known about "people" is generally derived from observations of men and about men, assuming that the male is the norm. Therefore, gender analysis often involves the need to focus upon women, but as part of a project that involves not just "adding women" but fundamentally reworking our understanding of personhood.

We have argued for the view that women, men, nature and environment are all culturally and historically specific, and cannot be disconnected from the context in which they are formed and reproduced. The concept of reproduction has been elaborated as one framework for gender analysis. The term reproduction has several levels and meanings:

- biological reproduction refers to the process of child bearing and rearing
- generational or daily reproduction refers to the maintenance of the domestic group (e.g. food production and processing, water collection, etc.)
- social reproduction involves a range of wider processes by which societies are reproduced over time (e.g. education socializes children and thereby assigns them, to some extent, to positions in divisions of labour).

In all these arenas women and men play different roles, and have different rights, responsibilities and expectations. The concept of reproduction is especially useful for understanding gender issues in environmental change, because it illuminates the gender differentials in fertility behaviour and in divisions of labour at the household level. The concept also links these into societal processes such as the reproduction and renegotiation of norms and values, the trajectories of accumulation and impoverishment, and the changing character of development policies and interventions. The concept of sustainability (see Ch. 2) is in a way analogous to environmental reproduction, which could be seen as one process of social reproduction.

Finally, before passing on to examine the evidence for differential impact of environmental degradation on men and women, let us examine the term gender relations. This signifies the importance of analyzing women and men in relation to each other, rather than in isolation. It does not mean that we necessarily devote equal attention to both genders, but it does mean that we focus upon the relations between people as bearers of gender identity. Gender relations exist at several levels. For example, between husbands and wives (as well as other kin categories) within households, between households, between employers and employees, and at societal level between the State and citizens.

Impact of environmental change on men and women

Gender divisions of labour

One important point to make in arguing that environmental degradation is a gendered process is that, since men and women occupy different positions in processes of production and reproduction, they will experience changes in the environment differently. Below, we consider the impact on gender divisions of responsibility, but here we focus upon absolute labour inputs of men and women and their relative proportions. Some examples will illustrate this, but first we can introduce a useful way of distinguishing different types of gender divisions of labour. Whitehead (1981) has used the terms "sex segregated" and "sex sequential" to describe agricultural production processes, and we can apply this to reproductive labour too.

A "sex segregated" division of labour is one in which the entire production process is largely controlled by either men or women. For example beer brewing in many rural African societies is "women's work" from beginning (wood collection and germination of grain) to end (sale or home consumption). A "sex sequential" division of labour, by contrast, is characterized by the interleaving of tasks mostly done by men and those mostly done by women. For example, in cattle-keeping, men may herd, plough with, and slaughter cattle, whereas women may collect fodder for stall-feeding the cattle, milk the cattle, process milk products and sell these.

The extent to which the negative impact of degradation on particular groups of people becomes an incentive towards conservation is partly dependent on this distinction. The degree to which environmental impacts are sex segregated helps to indicate the appropriate target group for development interventions and relevant forms of project or programme implementation, for example, gender targeted (at either men or women) or gender integrated.

Table 4.1 shows the different tasks, time spent and seasonal distribution of work done by men and women in Nepal. It demonstrates the large amount of women's time, relative both to men's labour inputs and to women's other activities such as agricultural production, that goes into the collection of wood, water and fodder. This same study found that with

Table 4.2 Seasonal pattern of time allocation in Nepal (1982/83).

Activity	April–June			July–September			October–December			January–March		
	M	W	C	M	W	C	M	W	C	M	W	C
Agricultural work (hours/person/day)												
Field work	2.2	2.1	0.1	4.1	3.4	0.0	3.8	3.4	0.0	2.3	2.1	0.1
Employment	0.5	0.2	–	0.6	0.2	–	0.5	0.0	–	1.6	0.1	–
Subtotal	2.7	2.3	0.1	4.7	3.6	0.0	4.3	3.4	0.0	3.9	2.2	0.1
Support activities												
Fuelwood collection	0.4	2.0	0.2	0.0	0.9	0.2	0.0	0.8	0.1	0.1	0.9	0.0
Water collection	0.2	1.6	0.3	0.0	0.9	0.4	0.1	0.9	0.1	0.1	1.2	0.1
Grass collection	1.2	0.9	0.3	0.1	2.4	0.7	0.1	0.4	0.1	0.0	0.2	0.0
Leaf fodder collection	0.1	0.3	0.0	0.0	0.0	0.0	0.2	0.4	0.1	0.1	0.7	0.0
Grazing	–	–	2.5	0.0	0.0	2.5	0.0	0.0	2.1	0.0	0.0	2.6
Food processing	0.2	0.7	–	0.2	0.7	–	0.2	0.7	–	0.2	0.7	–
Cooking	0.4	2.2	–	0.4	2.4	–	0.3	2.1	–	0.3	1.7	–
Subtotal	2.5	7.7	3.3	0.7	7.3	1.3	1.0	5.3	2.5	0.8	5.4	2.7
Total	5.2	10.0	3.4	5.4	10.9	1.3	5.3	8.7	2.5	4.7	7.6	2.8

Source: Kumar & Hotchkiss (1988).
Based on data collected by Nepal Agricultural Projects Service Center; the Food and Agriculture Organization of the United Nations; and the International Food Policy Research Institute, *Nepal energy and nutrition survey, 1982/83,* Western Region, Nepal.
M = men, W = women, C = children (6 to 15 years old are included).
Fieldwork was recorded by crop, which cuts across individual quarters. Therefore, it is aggregated into two semi-annual periods: dry-season crops were assigned the first and fourth quarters, and wet-season crops were assigned the second and third quarters.
Employment includes seasonal migration.
Since grazing of cattle was mostly performed by children and no data on individual grazing time are available, it is all assigned to that category.

deforestation the average increase in fuel collection time for women was 50–60 per cent (Kumar & Hotchkiss 1988). Although these figures need to be seen in the light of other adjustments to the farming systems consequent to deforestation, there is little doubt that women in many parts of the developing world bear the brunt of deforestation, since they are the primary fuel-wood collectors.

Since environmental degradation is responsible for changing the availability of water in rural areas, the existing gender divisions of labour in many parts of the Third World will mean that women are more affected by increasing distances to water sources than are men. Changing water availability may be caused by changing rainfall patterns, farming systems and water use, but whatever the cause there is a clear impact on women because of their daily reproductive labour. As well as the time expended by women in water collection, there are serious health effects of carrying heavy loads, which include the cranial depression and severe headaches of women who use head straps, spinal deformities, osteoarthritis of the knees, and also injuries. A study in a clinic in Bangladesh found that 50 per cent of broken-neck cases were the result of falls while carrying heavy loads (Curtis 1986).

The degree to which land degradation caused by human agency affects water availability is variable; in Asia, overexploitation of groundwater by tubewell extraction for irrigation may generate domestic water shortages that affect women, but the impact of, say, deforestation on water quantity is less clear. Deforestation does not necessarily reduce infiltration of rainfall, nor does increased runoff affect water tables in a predictable fashion. In one study, in the upper Sabi Valley of Zimbabwe, streamflow has been shown to increase, independently of rainfall, in periods when deforestation was rapid (du Toit 1985). In this area, women generally walk a distance of 4–8 km to collect water, mostly (59%) from surface-water sources, but it is not clear if, or how, the reported drying up of springs and pools is related to this. Anthropogenic degradation apart, when water supply is under stress, gender divisions of labour put women at the sharp end of water shortage. A study of water access during drought in Nkayi District, Zimbabwe, found that (mostly female) domestic users of borehole water had to give precedence at water points to (male) cattle watering (Elson & Cleaver 1994). Water quality degradation may also have a particular impact on women because of the gender divisions of labour. For example, women are usually responsible for transplanting in rice farming systems and, where fertiliser applications are high, women transplanters suffer from skin diseases related to long exposure to chemicals in the water. In Sri Lanka, rice transplanting is done by gangs of migrant women workers who have been found to suffer from such chemical damage (Jayatilaka 1990).

A large literature has documented the important role of women in agriculture and food production (e.g. Kandiyoti 1985), and many micro-studies of time use have revealed the differences in male and female labour inputs to crop husbandry in particular. For example, in rural Botswana, men spend 5.4 per cent of their total time on crop husbandry, whereas the corresponding figure for women is 8 per cent (Mueller 1984). Similarly, in Cameroon in 1974, women spent 300 hours per year on food crops, whereas men spent 50 hours (Henn 1978). However, Table 4.1 indicates that male inputs to agriculture in the western region of Nepal somewhat exceed those of women: gender stereotypes in which women are said to do all the farm work are misleading.

Gender divisions of labour in agriculture show that the impact of declining yields on arable land impinges differently upon the workloads of men and women, although it is not easy to predict the direction of such changes. Falling fertility in a land-constrained farming system may mean less work for women, as there may be less weeding to do and less crop to harvest and process. What gender analysis offers is not a set of generalizations about, for example, men's and women's contributions to farm production, but a means of rethinking, for any specific location, the implications of a gender-disaggregated account of farm production.

The importance of understanding gender divisions of labour is not only related to gauging the amount of work done by men and women, but also because of the need to recognize that men and women do different work. In the Botswana case cited above, men spent 19 per cent of their total time in animal husbandry, whereas women spent 2 per cent on this activity. Men in this gender division of labour may be expected to be more affected than women, at least in terms of labour time, by rangeland degradation. For example, cattle may need to be herded to grazing lands at increasing distances. In all these cases we need to be careful not to read too much into the static picture of the gender division of labour derived from survey research alone, which fails to capture the responses people make to environmental stress that alleviates negative effects. In the Botswanan example, men faced with declining grazing resources may evolve collective herding practices that cope with this stress without involving them in greater herding labour time.

To summarize, we can say that the gender division of labour, narrowly defined, shows us that men and women will be differently affected by different elements of environmental degradation, in terms of both the absolute labour time use and the relative contribution to any particular task by each gender. However, there are limits to what we can learn from the gender division of labour, and too often assumptions are made on the basis of this alone, assumptions that fail to recognize that labour and responsibility are not the same. The next section examines the significance of what we might call gender divisions of responsibility.

Gender divisions of responsibility

Domestic groups can be seen to operate on the basis of sets of implicit contracts between members, defining areas of rights and responsibilities, which guide behaviour and action. These are not immutable, indeed they are contested, and struggled over, and changed with time, and they may vary with class or other social divisions. However, they are an important element in mediating the impact of environmental degradation. We will look here at the "conjugal contract" (i.e. between husbands and wives) as one example of how the gender dimension of such contracts mediates in the process by which the impacts of environmental degradation are experienced as an incentive towards conservation. In a study of tree-use in Kenya, it was found that, in the past, landownership disputes were frequently resolved on the grounds of which of the parties to the dispute were able to claim the oldest and most mature trees on the farm, and that this continued to influence tree-planting behaviour of men and women:

> Despite the growing desperation of women in their search for new sources of firewood, they are still barred from planting trees on the farm for that purpose. Such an action would be considered almost an act of rebellion against the social order; in fact, many women would be too afraid of censure even to consider doing it openly. Moreover, because the provision of fuel wood has always been a woman's responsibility, it is generally felt that if she admitted that she could no longer cope, it would be tantamount to admitting failure as a wife.
>
> (Bradley 1991: 213)

This passage shows a conjugal contract that embodies the legitimacy of the husband's land rights and the illegitimacy of the wife's labour as a basis of land rights. It also shows the strength of the expectations that the wife should meet her obligations, and thereby internalize difficulties. The structural weakness of women *vis-à-vis* men in the conjugal contract is revealed in this example, for the terms of the contract protect the position of the husband from challenges by the wife. This illustrates the problem with using the term "contract", a term signifying equality, mutual advantage and voluntary co-operation. We should not jump to the conclusion that the entire conjugal contract embodies the subordination of women without an analysis of all the other elements of the arrangement, some of which may operate to women's advantage, and we should not imagine that such contracts are either fixed or adhered to by all individuals. However, they do constitute something of a societal view of gendered rights and responsibilities within marriage, and these may be utilized in bargaining, asserting, or disputing the viewpoint of the individual husband or wife. The impact of environmental degradation on men and women is filtered through these processes, and cannot be simply read off from the divisions of labour.

Gender divisions of income

In addition to gender divisions of labour and responsibilities, we also need to know about gender divisions in access to and control of incomes, both cash and kind, since men and women vary in both the acquisition and disposal of incomes. Although women make substantial labour contributions to farm production, the degree to which the outputs of farm production are controlled by women is frequently much more limited. Household food may be stored in granaries, which are controlled by men; decisions over crop sales, gifts and disposals may be male dominated; and marketing of cash crops may be a male prerogative. In such circumstances the degree to which men and women are motivated to either environmentally damaging behaviour (e.g. expansion of crop areas to grow a cash crop despite the exacerbation of soil erosion) or conservationist behaviour (e.g. the terracing of fields) will be influenced by the differential income access of each. This is not to suggest that women do not benefit from increased male income; they do, but this is mediated by the character of transfers within the household (culturally very variable) as well as by markets. One study of domestic budgets in the Cameroons found that:

> the welfare of the women's economy taken as a whole is positively responsive to levels of income in the male economy, but less through transfers into a collective pooled income than through the transactions of exchange and the market. (Guyer 1988: 169)

As well as joint incomes, of food and cash, we now know that, in most households, individuals also have individual incomes that they control privately. For example, among the Hausa of West Africa both men and women obtain such private incomes by a variety of means. Men farm private plots of land as individuals, they trade and they manufacture, whereas women produce snack foods for sale, do domestic and agricultural wage work, trade, make and sell craft goods, and collect wild produce for sale and charge for services (Jackson 1985). Access by men and women to such incomes is often more direct, being derived from what are frequently "sex-segregated" activities, and the consequences of particular resource degradations on men and women engaged in such activities may similarly be more direct. For example, pottery may be a form of off-farm enterprise that generates private incomes for rural women, and a scarcity of particular tree species for firing may impact directly and unambiguously upon women potters.

Many recent studies have shown differences in the expenditure patterns of men and women. The Cameroonian study by Guyer (1988: 165) estimated that women use three quarters of their personal incomes on food and routine household supplies, whereas men use about one quarter for this purpose, and Bruce (1989) reviews the accumulating evidence for gender differences in income expenditures as well as savings. In the disposal of incomes, the differential roles of men and women are an important element in both understanding the incentives towards conservation and the effective ability to participate in projects and programmes. Those that increase personal disposable incomes for particular individuals may well find it easier to identify a target group of participators than those that do not. Similarly, a programme such as the promotion of fuel-efficient stoves needs to take account of who has the ability as well as the incentives to purchase improved domestic equipment.

Access to and control of resources

Property rights are social relations. That is, they represent relationships between people and people, rather than people and things. For example, the title deed to a piece of land represents the acknowledgement by society of the legitimacy of an individual's claim to more or less exclusive control of that land. Thus, property rights are gendered in that gender identities enter into all social relations.

Men and women experience access to land in profoundly different ways in most rural Third World societies. Land may be accessed under a variety of different tenure arrangements, none of which is equally available to men and women. For example, in primary rights to land, patrilineal inheritance systems operate to exclude daughters; purchased land demands cash, which is generally less available to women (because of wage rate differentials, gender bias in employment, etc.) than men; sharecropping requires a source of unpaid (household) labour that cannot generally be commanded by women; and so on. Where women do have access to land, this is commonly in the form of secondary rights to land, that is those that are conditional upon a particular relationship, such as marriage. In these circumstances, women are able to farm household land for collective consumption or may be allocated land by husbands for personal use. Such access depends on the marriage remaining viable; divorce or widowhood can terminate it. Thus, although all property relations may be social, for women they are more social than for men. The significance of this for environmental degradation is that the benefits of land-improving conservation technologies are perceived differently by men and women, and women may quite rationally be indifferent to long-term strategies for land improvement. Land investment behaviour is related to security of land rights, and women lack permanent, inalienable rights.

Access to resources other than land is also very important. Women's environmental relations are also refracted through differential access to labour resources and to knowledge. Senior men are able to command the labour of other household members (wives and children) directly and junior men expect that, as they age, marry and reproduce, they will come to command the labour of others. For women a similar process may take place with ageing but on a modest scale. A mother may be able to make use of the labour of daughters and son's wives for domestic labour, but a father typically has both a wider range of kin and "dependants" to call upon, as well as the right to claim larger quantities of labour over a wider range of tasks. The problem of access to cash resources also inhibits women from making use of hired labour to the same extent as men. In sum, women are not able to control the labour of others to the same degree as men can. This is of great significance in labour-scarce economies, such as much of rural Africa, where pathways of economic accumulation may be based upon labour control. For policy and implementation, the appropriateness of conservation technologies with a high labour input is thrown into question by the particular severity of labour shortages faced by farm women.

Finally, the gender differences in environmental knowledges impinge upon how men and women manage natural resources and how they understand processes of degradation. As well as the variations in particular areas of knowledge, such as plant varieties or soil types, men and women relate to knowledge systems in different ways. A study in Zimbabwe, which examined the explanations for environmental change offered by local informants, reported that:

> Not everyone engages in this debate. Many say they do not know the reasons for, or fully understand environmental changes Because of the very commonly held association between trees and rainfall and the link to land ancestral spirit guardianship, some say the knowledge is specialized and defer to these sources. Women often defer to the older men on this issue. (McGregor 1991: 129)

A footnote remarks that women can attend ancestral spirit ceremonies only after the menopause and "frequently seemed more cynical than the men". Old women in this study were particularly concerned about the drying out of springs, and explained this with reference to a variety of causes, including pollution by the use of "salty" carrying jars, such as those that had been used on fires. Clearly, women were excluded from some areas of environmental knowledge and debate, were more concerned about some environmental changes than others, and were inclined to different forms of explanation. If unrealistic expectations are to be avoided, environmental interventions need to anticipate both the varied content of the knowledge held by men and women, and also the gender politics of knowledge.

Kinship and domestic groups

One of the areas of gender analysis that illuminates both the processes of environmental change and conservation interventions is that of gender relations at the household level, which we discuss here with a focus upon kinship and intra-household relations.

The debate about the nature of the household that has developed from feminist scholarship has significant implications for environmentalist thinking and practice, as well as for development more broadly (Guyer 1981). The household has been conceived as a unitary body with a range of functions – production, consumption, residence, reproduction, and so on – in concepts and models, as well as in descriptive empirical work, development policy and practice. In recent years this has been challenged by alternative views of households as having different forms and functions, according to class and other social divisions (Netting et al. 1984), as well as stages in the development cycle of the household (Goody 1958); that is, in the processes of household formation, expansion and dissolution. However, the critique of the unitary nature of households has been made most strongly from the perspective of the conflicting interests of men and women within households.

From our discussion of gender divisions of labour, rights and responsibilities, we can see some of the ways in which men and women may have distinct and different interests, although they are members of the same household. The question of how differing self-interests are reflected in decision-making is important, and it raises directly the issue of power in decision-making. Economic models have sought to overcome the problem of how individual interests of household members can be aggregated to a joint interest by assuming the existence of a benevolent dictator who makes decisions in the interests of the household as a whole (Ellis 1988). It has not been difficult for gender analysts to show that household heads (predominantly male) do not control all decision-making, nor do they necessarily arbitrate fairly in the collective interest, and that gender conflict of interests and outcomes are present to varying degrees in most, if not all, societies (Folbre 1986, Bruce 1989). How then does decision-making take place in the context of potentially conflicting interests of household members?

The "power" of individual household members to determine the outcome of decision-making processes is gender differentiated, although it would be misleading to suggest that women are always overruled by men. Although the outcomes are indeterminate, Amartya Sen (1990) suggests that the objective positions of men and women in the event of marital breakdown and the subjective perceptions (of self-interest, of self worth and of labour value, for example) of household members are important factors in the "co-operative conflict" that characterizes household interaction. Thus, where divorced women are especially poor and their social status is low, where women are socialized to prioritize the interests of others (e.g. children), where they subscribe to gender ideologies that denigrate women, and where women's labour contribution to household maintenance (e.g. domestic work) is devalued – then it is likely that household women will be in a weak position to influence decision-making, and outcomes may well reflect male interests. Of course, this is not always the case, and men may have to pay a price for ensuring the co-operation of household women in endeavours that are not in their interests. A gender analysis of decision-making draws upon an understanding of the rights and position of women in society at large (including those not in conjugal relationships), a knowledge of gender ideologies (including religion) and the study of perceptions of the value of work.

Any action to mitigate the impact of environmental degradation or to catalyze conservation activities or the adoption of conservation technologies will depend upon an understanding of household decision-making. Why do households allocate their resources and relate to their environments in the way they do, and how could this be changed? This question cannot be answered without examining the gendered processes of household decision-making.

There are many ways in which kinship, marriage and domestic development-cycles affect men and women differently in terms of environmental relations. The stylized (for exposition) account that follows is not intended to suggest that these social institutions give rise to a uniformity of effect or an inevitability of outcomes. Descent systems (e.g. patriliny or matriliny: traced through males or females) are sets of rules that regulate inheritance and social placement, and they colour environmental relations of men and women. In patrilineal societies, women are frequently defined in relation to men. In such societies, women are largely excluded from inheriting resources, and also probably from formal social significance in ritual and religion. Some consequences of this may be that they are alienated from resource control, and marginalized in social politics. Why should such an alienated and underpowered group be expected to have the same level of commitment to natural resources and the same time preferences as men?

Marriage is a variable institution, but it is probably reasonable to generalize that most marriage is patrilocal (i.e. the wife goes to live, at least initially, with her husbands father's household) and exogamous (i.e. marriage rules require unions to be between rather than within social groups) and therefore, upon marriage, women move to other households in other places. Women are therefore said in Hausaland (northern Nigeria) to be "strangers" in their marital homes. In Zimbabwe, McGregor notes, with regard to taboos governing tree-use, that many women burn some of the regulated species and that they that they do not know about the restrictions or that, as outsiders, such restrictions not apply to them (McGregor 1991). Marriage patterns make many women "outsiders", with consequences for environmental attitudes.

Life-cycles of men and women follow distinctly different patterns, which is not to suggest that all women follow the same pattern, but simply that kinship and marriage norms impose different templates upon the lives of men and women. Women relocate on marriage, on divorce and often on widowhood too, whereas in many societies men do not. The biological reproductive responsibilities of women are strongly age-related for women. For example, the experience of time-constraints by women is likely to be most acute for those aged between 20 and 35, for whom small childcare is onerous, and pressure towards labour-saving (even if environmentally damaging) livelihood strategies is acute. The dynamics of social reproduction and the gender differentials in roles are key elements in understanding environmental degradation and change.

Conclusion

Gender analysis tells us about some important structures that pattern the ways in which women and men relate to their environments. However, it is also about the struggles, explicit and implicit, to change those structures. One of the strengths of the reproduction concept is emphasis on change.

Divisions of labour change over time, as the bargaining processes within households discussed above begin to have different outcomes; for instance, as deforestation deepens, women may negotiate assistance from men with fuelwood collection. Also, divisions of income control may change if women become successfully involved with off-farm employment, or increasingly involved with wage labour where Green Revolution crop technologies are widely adopted. Gender differentials in terms of access to resources may be reconstituted in circumstances where women's groups have lobbied successfully for changing land laws or where credit becomes widely available to women. Gender divisions of responsibilities, as embodied in the conjugal contract, may change as marriage practices are transformed by women's agency. For example, by their non-compliance with bride-wealth exchanges on marriage, Zimbabwean women establish their greater independence and, incrementally, renegotiate the terms of the conjugal contract in their favour, with consequences for household resource management. The sets of rules and the institutions that men and women confront in their daily lives are not unchangeable. Some are more enduring than others, but none is beyond the reach of human agency.

Although we have concentrated here on explaining some elements of a gender analysis applied to environmental issues, this is only an illustrative and not an exhaustive treatment. There are important areas that have not been discussed here (e.g. the gender analysis of environmental relations of inter-household and community groups), but the concepts retain their relevance across a range of levels of analysis.

In this chapter, I have tried to show the value of an analytical framework as an approach to understanding gender relations and environmental change, rather than offer a futile attempt to summarize and generalize from the diverse and contradictory discourses on women and environment in development. Outcomes cannot be generalized across all cultures, and the experience of environmental degradation and conservation by men and women has to be reconsidered constantly. Conclusions may not always be easy to come to, but the exposure of gender trade-offs, the better to make informed choices, is in itself a useful exercise.

We cannot understand either the causes or the varying consequences of environmental degradation without a recognition that rural people are differentiated in ways that significantly alter the costs and benefits of environmental change for not only rich and poor but also men and women. A feminist social science has contributed a fundamental rethinking of how we understand science, of what we mean by concepts such as environment and conservation, and of how gender identities and relations are formed, reproduced and changed over time. In a more applied sense, gender analysis within the context of development supplies us with useful approaches to deploy towards improved understandings of men and women's environmental relations and more effective development interventions.

References

Agarwal, B. 1991. *Engendering the environment debate: lessons from the Indian subcontinent.* Distinguished Speaker Series 8, Centre for Advanced Study of International Development, Michigan State University.

Behnke, R. & I. Scoones 1991. *Rethinking range ecology: implications for rangeland management in Africa.* Drylands Network Paper and International Institute for Environment and Development Paper 33, Overseas Development Institute, London.

Beinhart, W. 1984. Soil erosion, conservationism and ideas about development: a southern African exploration. *Journal of Southern African Studies* 11, 52–83.

Biot, Y., R. Lambert, S. Perkins. 1992. *What's the problem? An essay on land degradation, science and development in sub-Saharan Africa.* Discussion Paper 222, School of Development Studies, University of East Anglia.

Blaikie P. 1986. *The political economy of soil erosion in developing countries.* Harlow: Longman.

Blaikie, P. & H. Brookfield. 1987. *Land degradation and society.* London: Methuen.

Bradley, P. N. 1991. *Woodfuel, women and woodlots,* vol. 1: *The foundations of a woodfuel strategy for East Africa.* London: Macmillan.

Bruce, J. 1989. Homes divided. *World Development* 17(7), 979–91.

Buchner, G., J. C. Burgess, V. Drake, T. Gameson, D. Hanrahan 1991. *Gender, environmental degradation and development: the extent of the problem.* Discussion Paper 91–04, London Environmental Economics Centre, International Institute for Environment and Development, London.

Cliffe, L. & R. Moorsom. 1979. Rural class formation and ecological collapse in Botswana. *Review of African Political Economy* 14–16, 35–52.

Curtis, V. 1986. *Women and the transport of water.* London: Intermediate Technology Publications.

Drinkwater, M. 1989a. *The state and agrarian change in Zimbabwe's communal areas: an application of critical theory.* PhD thesis, School of Development Studies, University of East Anglia.

— 1989b. Technical development and peasant impoverishment: land use policy in Zimbabwe's Midlands Province. *Journal of Southern African Studies* 15, 287–305.

du Toit, R. F. 1985. Soil loss, hydrological changes and conservation attitudes in the Sabi Catchment area of Zimbabwe. *EnvironmentalConservation* 12, 157–66.

Ellis, F. 1988. *Peasant economics: farm households and agrarian development.* Cambridge: Cambridge University Press.

Elson, D. & F. Cleaver. 1994. Gender and water resource management; integrating or marginalising women? Paper to OECD/DAC Expert Group on Gender and Water Resource Management, Stockholm 1–3 December 1993.

Folbre, N. 1986. Hearts and spades: paradigms of household economics. *World*

Development 14, 245–255.

Goody, J. (ed.) 1958. *The development cycle in domestic groups*. Cambridge: Cambridge University Press.

Guyer, J. 1981. Household and community in African studies. *African Studies Review* 24, 87–138.

— 1988. Dynamic approaches to domestic budgeting: cases and methods from Africa. In *A home divided: women and income in the Third World*, D. Dwyer & J. Bruce (eds), 155–72. Stanford: Stanford University Press.

Harlow, E. 1992. The human face of nature: environmental values and the limits of non-anthropocentrism. *Environmental Ethics* 14, 27–42.

Henn, J. 1978. *Peasants, workers and capital: the political economy of labour and incomes in Cameroon*. PhD thesis, Department of Anthropology, Harvard University.

Jackson, C. 1992. *Gender, women and environment: harmony or discord?*. Discussion Paper GA1D6, School of Development Studies, University of East Anglia.

— 1985. *The Kano River Project*. West Hartford, Connecticut: Kumarian Press.

Jayatilaka, W. 1990. *Seasonal migrant female workers who transplant paddy*. Research report, The Women's Bureau, Ministry of Teaching Hospitals and Women's Affairs, Sri Lanka.

Jodha, N. S. 1986. Common property resources and the rural poor. *Economic and Political Weekly* 21(27), 1169–81.

Kandiyoti, D. 1985. *Women in rural production systems: problems and policies*. Paris: UNESCO.

Kumar, S. K. & D. Hotchkiss 1988. *Consequences of deforestation for women's time allocation, agricultural production, and nutrition in hill areas of Nepal*. Research Report 69, Food Policy Research Institute, Washington DC.

Levi-Strauss, C. 1969. *The elementary structures of kinship*. Boston: Beacon Press.

MacCormack, C. & M. Strathern (eds) 1980. *Nature, culture and gender*. Cambridge: Cambridge University Press.

MacKenzie, J. 1988. *The empire of nature: hunting, conservationism and British imperialism*. Manchester: Manchester University Press.

McGregor, J. 1991. *Woodland resources: ecology policy and ideology. An historical case study of woodland use in Shurugwi Communal Area, Zimbabwe*. PhD thesis, Department of Geography, Loughborough University.

Mearns, R. 1991. *Environmental implications of structural adjustment: reflections on the scientific method*. Discussion Paper 284, Institute of Development Studies, Falmer, Sussex.

Merchant, C. 1980. *The death of nature: women, ecology and the scientific revolution*. New York: Harper & Row.

Mueller, E. 1984. The value and allocation of time in rural Botswana. *Journal of Development Economics* 15, 329–60.

Netting, R., R. Wilk, E. Arnould (eds) 1984. *Households: comparative and historical studies of the domestic group*. Berkeley: University of California Press.

Newby, H. 1980. *Green and pleasant land?* Harmondsworth: Penguin.

Pepper, D. 1984. *The roots of modern environmentalism*. London: Croom Helm.

Redclift, M. 1984. *Development and the environmental crisis: red or green alternatives?* London: Methuen.

Richards, P. 1987. *Philosophy and sociology of science: an introduction*. Oxford: Basil Blackwell.

Rogers, B. 1980. *The domestication of women: discrimination in developing societies*.

London: Tavistock.

Ruether, R. 1979. *New woman, new Earth*. New York: Seabury Press.

Sayers, J. 1982. *Biological politics: feminist and anti-feminist perspectives*. London: Tavistock.

Sen, A. 1990. Gender and cooperative conflicts. In *Persistent inequalities: women and world development*, I. Tinker (ed.), 123–49. New York: Oxford University Press.

Shiva, V. 1988. *Staying alive: women, ecology and survival*. London: Zed.

Strathern, M. 1980. No nature, no culture: the Hagen case. In *Nature, culture and gender*, C. MacCormack & M. Strathern (eds), 174–223. Cambridge: Cambridge University Press.

Stretton, H. 1986. *Capitalism, socialism and the environment*. London: Cambridge University Press.

Talle, A. 1988. *Women at a loss: changes in Maasai pastoralism and their effects on gender relations*. Stockholm: University of Stockholm Press.

Thompson, M. & M. Warburton 1985. Uncertainty on a Himalayan scale. *Mountain Research and Development* 5, 3–34.

White, L. 1967. The historical roots of our ecologic crisis. *Science* 155, 1203–207.

Whitehead, A. 1984. Men and women, kinship and property: some general issues. In *Women and property, women as property*, R. Hirschon (ed.), 176–95. London: Croom Helm.

— 1990. Food crisis and African conflict in the African countryside. In *The food question; profits versus people?*, H. Bernstein, B. Crow, M. Mackintosh, C. Martin (eds), 54–68. London: Earthscan.

— 1981. *A conceptual framework for the analysis of the effects of technological change on rural women*. World Employment Programme Working Paper 79, International Labour Organisation, Geneva.

CHAPTER FIVE

Biotechnology: a servant of development?

Stephen Morse

Editors' introduction

The biodiversity debates at the Earth Summit in Rio (June 1992) provided for many one of those rare moments that encapsulate the reasons why we continue to have great inequality between people on this planet. At the summit President George Bush refused to sign the framework Convention on Biodiversity on the grounds that it could be detrimental to the US economy. The treaty was a deal by which developed countries could access biological resources (such as genes) in return for money and technical assistance. President Bush's refusal to sign the treaty provided one of the news highlights of the conference, and brought into sharp focus the conflict between national interest and wider development. It also highlighted the wealth-creating potential of such biological resources.

Many look to science as the provider of answers to problems that threaten to engulf us. AIDS and global warming (Ch. 3) are two specific and very different threats that were highlighted in the 1980s, and to which science has, as yet, no immediate fixes. Instead science has helped in an understanding of the problems and in doing so has provided the basis for suggested action. However, although a disease such as AIDS can be targeted for attention, how does one target a problem such as poverty, which has a much more complex set of causations? One of the answers being put forward by many is the new and exciting field of biotechnology. The Agenda 21 document that came out of the Rio conference provided a sober analysis of the contribution that biotechnology could make to development:

> By itself, biotechnology cannot resolve all the fundamental problems of environment and development, so expectations need to be tempered by realism. Nevertheless, it promises to make a significant contribution in enabling the development of, for example, better health care, enhanced food security through sustainable agricultural practices,

131

improved supplies of potable water, more efficient industrial development processes for transforming raw materials, support for sustainable
methods of afforestation and reforestation, and detoxification of hazardous wastes.   *Agenda 21*, Chapter 16, paragraph 1

There can be few areas of recent scientific development that have received
as much debate as biotechnology, and in particular genetic engineering.
Many authors have also speculated about the potential benefits, such as
those listed above, and indeed threats to developing countries that may result
from the application of biotechnology. Some of these arguments impinge
upon the more general debates over biodiversity highlighted by Stocking at
al. (Ch. 6), and agricultural sustainability summarized by Gibbon at al. (Ch.
2). Agenda 21 summarized the challenges for the post-Rio future:

Biotechnology also offers new opportunities for global partnerships,
especially between the countries rich in biological resources (which
include genetic resources) but lacking the expertise and investments
needed to apply such resources through biotechnology and the countries that have developed the technological expertise to transform
biological resources so that they serve the needs of sustainable development.   *Agenda 21*, Chapter 16, paragraph 1

What is biotechnology?

Biotechnology has recently received a great deal of media attention. Its
promise to revolutionize our lives, and the potential dangers that may be
inherent in this revolution, have often been repeated. Some of its proponents have even seen it as the ultimate "technical fix" in development – a
means by which all humankind will reach Utopia. In this chapter I will
examine biotechnology within a development context, by looking at the
science and how it potentially could assist (or hinder) the process of development in developing countries. Such a discussion is particularly opportune, as at the time of writing the first commercial release of a genetically
engineered product has recently happened. Indeed, lessons and mistakes
drawn from experience may quickly supersede hypothetical discussion.

Definitions of biotechnology abound, largely because there is some disa-
greement as to what the term covers (Table 5.1). One definition even tries
to please everyone by referring to biotechnology as the utilization of biological systems to yield products useful to humankind, a definition that takes
in the whole of agriculture and fermentation technology. Indeed, if one
reasonably regards a human being as a "biological system", the definition

Table 5.1 Some definitions of biotechnology.

– The application of scientific and engineering principles to the processing of materials by biological agents to provide goods and services. (Bull et al., cited in Farrington & Greeley 1989)

– The application of biological science to the manipulation and use of living things for human ends. (Walgate 1990)

– Any technique that uses living organisms or substances from those organisms to make or modify a product, to improve plants or animals, or to develop micro-organisms for specific uses. (Persley 1990)

covers virtually all of industry. A second problem for the non-specialist is that the technologies that are often included under the umbrella of biotechnology are diverse and complex (Table 5.2), and typically require a good working knowledge of molecular biology and biochemistry to understand and appreciate fully. However, for many people the terms "biotechnology" and "genetic engineering" have become synonymous, and authors frequently use the two terms interchangeably. This is probably attributable to the more powerful and radical nature of genetic engineering when compared with the other biotechnologies.

For the purposes of this chapter, the emphasis will be on the use of "genetic engineering" along with related and associated techniques in agriculture, and the term "biotechnology" will be used to cover the range of processes and technologies set out in Table 5.2. Emphasis will also be placed on crops, although similar issues relate to animal husbandry.

Table 5.2 Some biotechnologies.

Technology name	Function
Genetic engineering Chromosome engineering Protoplast fusion	Manipulation of genetic material
Gene markers	Use of molecular markers for specific characteristics
Tissue culture	Generation of whole plants from sections of plant tissue
Diagnostic techniques	New techniques for the detection of extremely small quantities of a chemical
Fermentation technology	Generation of useful products from fermentation vessels

The above are typically combined. For example, a bacterium may be subjected to genetic engineering in order to include a useful gene from another organism into its genome. The bacterium may be cultivated in a fermentation vessel, and the useful product generated from the new gene extracted.

There is currently a wealth of books, papers and articles on biotechnology and its potential impact on developing countries; some of the more recent publications have been listed at the end of this chapter. Given this material, the reader is entitled to ask about the need for yet another addition, especially one that can only occupy a few pages. However, development is a great deal more complex than biotechnology, a point that may not always be apparent to, or even accepted by, biotechnologists, and I believes that much of the published material often fails to reflect this complexity.

Indeed, placing biotechnology firmly within a development context is one of the aims of this chapter, and its inclusion within a book that covers a wide range of important issues in development will help to achieve that aim. Secondly, given that the thrust of this book is to examine development issues within a global context, biotechnology will be viewed very much in the same mode. In particular, some of the issues that are commonly mentioned in this debate will be dissected from a pragmatic viewpoint.

Given that, at the time of writing, the products of some biotechnologies, particularly genetic engineering, have not yet been employed on a wide scale in developed, let alone developing, countries, much of the literature on potential impacts is based on speculation. Experiences, for example with conventional plant breeding techniques, can be drawn upon to develop some reasonable conclusions, but the truth is simply that no-one knows for sure what will happen. The contribution provided by this chapter is no different, and the third aim of this chapter is to provide a reader with some of the basic ideas and arguments to provoke some thought and discussion.

Genetic manipulation: a new idea?

Genetic engineering implies a human-mediated alteration in the genetic make-up of an organism. Although this may sound very new and "high tech", human-mediated changes in the genetic structure of plants and animals have been around for a long time and are not inherently high tech. Most if not all the common crop varieties and animal breeds now used in agriculture are the result of many years of selective breeding. Indeed, many of the common crop plants grown today, such as wheat and maize, are far removed genetically from their wild ancestors. However, such examples have relied almost entirely upon cross fertilization within a species or between individuals of closely related species. Also, the crosses have been by way of the organism's own natural mechanisms of reproduction. Indeed, to this day, production of new crop varieties relies largely upon the transfer of pollen from one line to another, followed by a painstaking search among the offspring for plants that have the desired characteristics.

Before the nineteenth century, plant and animal breeding were practised with little idea as to the mechanisms involved in the inheritance of characteristics. Trial and error showed that, if two distinct lines were crossed, the offspring had a mixture of characteristics inherited from the parents. In the nineteenth century, an Austrian monk named Gregor Mendel formalized some of the basic laws of genetics, and progress became very rapid. The twentieth century has seen the discovery of the structure of the nucleic acids (DNA and RNA) the chemical group at the very heart of each gene from all life on Earth, and the development of an understanding as to how this structure can influence the characteristics of living organisms.

The aims of a plant breeding programme can be diverse, and can include many objectives of immediate relevance to agricultural development and resource-poor farming. Targets such as increased yield, pest and disease resistance, salt tolerance, drought tolerance, early maturity, plant stature, increased nutritional value, increased storability and increased cookability – are examples of aims that may be attractive to small-scale farmers in vulnerable environments. The methods involved are relatively straightforward, especially if the characteristic is linked to a single gene, although time-consuming and expensive. The major problems include finding useful genes in the germplasm and ensuring that a gene has been transferred and is producing the desired characteristic in the offspring. This may be relatively easy with a recognizable character such as seed colour, but not so easy if the character is expressed only under certain circumstances such as the presence of a pest or disease. Plant breeders often look for a genetic marker, typically a clearly distinguishable characteristic, which helps indicate whether a useful gene has been transferred. Some new techniques take this one stage further by allowing the detection at the molecular level of a marker for the characteristic (Arus & Moreno-Gonzalez 1993).

However, before a gene can be transferred into a line using conventional techniques, it has to exist somewhere within either the available germplasm of the species or a close relative with which it can cross. This fundamental limitation restricts the range of characteristics that can be incorporated with these techniques into a crop species. Plant breeders have developed some methods (for example, the use of "bridging species") that allow gene transfer between more distantly related species, but the applicability of these is limited. In effect there is a natural barrier that severely restricts the scope of conventional plant breeding.

Further difficulties arise if the character is produced by several genes, not just one (a polygenic or multigenic character). Transfer of the desired characteristic would require transfer of all the relevant genes. Again, the use of molecular markers and gene maps could help plant breeders work with such multigenic characters.

Conventional plant breeding has contributed many new crop varieties to

135

agriculture, and these now account for much of the cultivated area on all continents (Lipton 1989). Most of the international agricultural research centres produce new varieties as a central component of their research strategy; indeed, each has a mandate to do this for a particular group of crops. With all this effort, and money, have the new crop varieties really helped small farmers in developing countries?

Are new crop varieties beneficial for resource-poor farmers?

New crop varieties have some important advantages, besides the purely agronomic, from the viewpoint of a resource-poor farmer. Farmers are familiar with different crop varieties, and are well able to adapt them to their situation. A crop may have many different varieties, each suitable for a particular use or environment, and, although often classified under the generic term "local", many of the varieties currently cultivated may be introductions from 10, 20, 30 or more years ago. Farmers are excellent experimenters (Richards 1989), and have no difficulty determining whether a new variety offers something useful, and, if so, how the variety could be best exploited. That most crops are not native to the area in which they are grown attests to the abilities of generations of farmers to exploit the potential of new genetic material. Maize and cassava, for example, although now commonly cultivated throughout Africa and often referred to as "traditional crops", were introduced into that continent over only the past few hundred years (Anthony et al. 1979).

The second major advantage of new crop varieties rests on the fact that they can be easily replicated locally at a relatively low cost. This facilitates the involvement of non-government organizations (NGOs) as well as farmers in seed multiplication and distribution (Cromwell & Wiggins 1993). Of course, this is not true for hybrid varieties that have to be bought every year. Other technologies frequently emphasized in development programmes, such as artificial fertilizer, pesticides and machinery, have high costs of purchase, storage, maintenance and distribution.

A third advantage of new crop varieties rests with their minimal negative impact upon the environment. Resistance to pests and diseases is obviously of particular relevance in this respect. Again, this is in sharp contrast to the oft-quoted damaging impacts of fertilizers and pesticides. In addition, plant resistance to insect pests can perhaps be integrated with low dose rates of insecticide or even biological control (Boethel & Eikenbary 1986). The level of control achieved with the combination of strategies is often greater than with either on its own.

The above advantages of new crop varieties have been partly responsible for their prominence in many research and extension programmes. However, they do have disadvantages, some of which have been recently highlighted by Lipton (1989) and summarized in Table 5.3. In addition, there are the ubiquitous problems of cost and availability of skilled personnel required to develop the varieties. Of particular interest is the interaction between new varieties and input use. It has long been argued that, in general, new crop varieties require high levels of expensive, and environmentally damaging inputs, to realize their full potential (Reijntjes et al. 1992). This results in two serious problems. First, farmers become locked into a

Table 5.3 Some commonly quoted disadvantages of new crop varieties (partly after Lipton 1989, and others). The list is by no means complete, and the points do not apply to all new crop varieties

1	New varieties require more input (fertilizer, pesticide) use in order to out-perform "local" varieties. If used without inputs they may either perform worse than the "locals" or be of only marginal benefit.
2	New varieties are more sensitive to environmental stress (drought, pests, diseases etc.) than the "locals".
3	New varieties can't compete with the prevalent weed complex as well as the "local" varieties.
4	New varieties have a poorer taste and lower nutritional value than the "locals". They may also be harder to store or process. The result of these may be a lower market value of new varieties relative to the locals.
5	New varieties can't be "slotted" into the common cropping systems (e.g. intercrops) of many developing countries.
6	New varieties force farmers into using sole cropping systems. This leads to more use of expensive, and environmentally damaging, inputs.
7	The dependence on the use of expensive inputs to obtain the best yields from new varieties favours rich farmers at the expense of poor farmers. The result is an exacerbation of an existing social imbalance (poor farmers get poorer while rich farmers get richer).
8	New varieties help exacerbate gender imbalances by requiring more labour input from women.
9	New "high yielding" varieties drain more nutrients from the soil. This leads to resource degradation.
10	Use of "fast maturing" varieties leads to more crops in a season, and encourages the use of machines which in turn causes a displacement of labour.
11	Popularity of new varieties encourages genetic uniformity over an area. This leads to a loss of genetic diversity (shrinking of the gene pool), and results in increased instability of the system.
12	Resistance to pests and diseases which has been included in new crop varieties can "break down" over time.

spiral of increasing input use. Secondly, the rich farmers who can afford the inputs benefit most, whereas poor farmers usually miss out entirely because they cannot invest in the innovation.

However, perhaps some of the disadvantages listed in Table 5.3 may not be as serious in practice. For example, as van Emden (1987) has pointed out, there are few examples of resistance to insect pests breaking down, although the phenomenon has been commonly observed with resistance to many plant pathogenic fungi. Similarly, in sharp conflict with established "wisdom" based on extrapolation from experience gained with rice and wheat in the Green Revolution, some research has shown that new varieties of many crops may not necessarily require expensive inputs to out-yield local varieties (Morse & McNamara 1994). There is also much evidence of farmers being able to incorporate new varieties into existing low-input systems, including those based on intercropping (Morse & McNamara 1994).

Genetic engineering: the hype?

Given the long history of human involvement in altering the numbers and types of genes in plants and animals, why the "hype and hysteria" over genetic engineering? The genetic material for all life on Earth is based on a single group of chemicals, the nucleic acids, and a gene is a length of this chemical. Although the gross structure of nucleic acid is essentially the same for all species, variation in the fine detail of the structure produces the vast spectrum of life that we observe. Each species can be regarded as a unique set of genes, copies of which exist in each individual of that species, albeit with minor variation in the structure and number of some genes from one individual to another. Biologists illustratively refer to this set of genes from all the individuals of a species as the "gene pool". The frequency of individual genes (including mutations) in this pool is dependent upon natural or artificial selection operating upon the physical and behavioural characteristics the genes help produce (Dawkins 1986).

Individuals of each species possess barriers, the nature of which varies greatly from species to species, that normally prevent genes from other species "invading" the gene pool. In the wild, such invasions would, for the most part, harm the competitive ability of individuals and severely handicap them in the stark "arms race" forced by natural selection. Genetic engineering bypasses the natural barriers that prevent such invasions, allowing the transfer of genes between completely unrelated species, even between animals and bacteria. However, only beneficial transfers are selected by the genetic engineer, and the care and attention lavished by humans helps the "transgenic" individuals survive and prosper.

The potential of genetic engineering, with its ability to break the natural barriers for genetic transfer between species, is huge (Persley 1990, Fincham & Ravetz 1991), and this partly explains the hype. Genes coding for human insulin can be transferred to bacteria, allowing them to produce the chemical from simple nutrients. If a wild plant species has genes that code for resistance to a particular pest, these can be transferred to important crops. In effect, genetic engineering widens the source of genes available for incorporation into a useful organism. Indeed, ultimately, the potential is there for genes to be artificially mutated or even custom-made in a laboratory to supply a characteristic that does not exist anywhere in nature.

Besides an extensive broadening of the genetic resource base that can be used in a breeding programme, the use of genetic engineering allows a more targeted approach to gene transfer between organisms. Conventional crossing between plants of two varieties will result in the offspring having a 50:50 combination of genes from both parents. However, plant breeders typically wish to transfer perhaps only one gene into a variety that has many other useful characteristics. Therefore, the vast majority of genes (perhaps 99.99%) from one of the parents are not required. The transfer of just a few genes can be achieved with conventional plant breeding, but is time and labour consuming. Genetic engineering provides a much more precise and hence faster way of transferring these useful genes without the others. However, to be of any practical benefit this does assume that the other genes present in the target variety are of good quality. In other words, simply transferring one or two genes into a variety that generally performs poorly in the field may have little benefit, unless the transferred genes specifically correct the reason for the poor performance. This implies that a conventional plant breeding programme will need to be in place to take full advantage of genetic engineering.

Genetic engineering: the hysteria?

The "hysteria" associated with genetic engineering has essentially come from five directions. First, there is a basic moral question whether such gene transfers between species should be practised. Secondly, there are the potential problems of the "engineered" organism getting out of control, once released into the environment, or perhaps the genes could spread to other species, thereby producing new diseases, super weeds and similar dangers. Thirdly, and ironically, there is the cost of success. The danger exists that, if one or two engineered crop varieties are popular for whatever reason, many farmers will grow them and the diversity of genes in the gene pool will shrink. Fourthly, there are wider issues, such as who controls the new technology and its products, and what will be the effects on people's livelihoods. Finally, with development in mind, there is a view that genetic engineering is just another high-profile technical fix that does not really tackle the root causes of underdevelopment.

The question over the morality of genetic engineering is very different to the others mentioned above, in that it presents a *fait accompli*. Whether genetic engineering is moral, immoral or amoral cannot be argued on a scientific basis. Staugham (1992) refers to the moral argument against genetic engineering as an "intrinsic concern", whereas arguments based on environmental damage, loss of biodiversity, and so on are "extrinsic concerns". As pointed out by Staugham (ibid.), the moral issue is a complex one with many facets. It is also doubtful whether proponents on either side of this argument will ever be able to convince their opponents. For example, defendants of genetic engineering stress that humans have been involved in the genetic manipulation of plants and animals for thousands of years. Antagonists of genetic engineering, on the other hand, point out that none of these manipulations have involved the transfer of genes across a wide species barrier.

The moral question has been especially asked of genetic engineering in animals (Fox 1990, Linzey 1990, Ryder 1990) and humans. Indeed, the arguments become almost exponentially more intense as one moves from bacteria and viruses through to plants, animals and finally humans. In the latter case, almost any type of genetic interference, even gene mapping, is typically deemed unacceptable (Rogers 1990), with eugenics (improving human populations by selective breeding) almost seen as an inevitable consequence unless we are very careful (Gros 1989). Indeed, the intensity of the contemporary arguments on the morality of genetic interference could almost be used to re-establish and calibrate the ancient "chain of being" hierarchy (Pepper 1984), which ranked life in terms of its proximity to God!

The environmental damage argument resulting from engineered organisms going "out of control" in a natural environment is one that can be scientifically studied. Krimsky (1985) provides a fascinating insight into the social history of this research. Here it is suggested that organisms that have received genes from other species can become a nuisance, or possibly pass on its genes to other species that may in turn become a nuisance (Walgate 1990). The first problem is certainly one that is possible, as many examples of species introduction into new habitats have shown in the past. It may be argued that relatively few introductions have led to major environmental problems, but the potential is certainly there and cannot be avoided.

The second problem, that of genes "escaping" from an engineered organism has received much attention. For example, experiments have shown that marker genes were transferred between cultivated and wild radishes. Similarly, such transfers of "new" genetic material could also occur between a widely cultivated plant, oilseed rape, and other brassicas. These fears become especially crystallized when one considers that a major activity in genetic engineering at present is the transfer of herbicide resistance genes into crop plants (Kloppenburg 1988, Anon 1991, Fincham & Ravetz 1991, Hobbelink 1991). Indeed, the herbicide resistance issue is unusual in that both the biotechnology industry and its antagonists use environmental concerns as a mainstay for their arguments!

Although it is possible for genes to "escape" from an engineered organism, it should be remembered that gene transfer is restricted to closely related species. Indeed, one could argue that, if it was so easy to pass genes between species, then human mediated genetic engineering would not be necessary! Elaborate barriers have evolved over millennia to prevent such interspecies transfer of genes, and genetic engineering has to go to great lengths to enforce it. In situations where there is a risk of gene transfer, one answer may be introduce male sterility into the engineered plant, thereby preventing it from crossing with wild relatives. However, this solution would require the farmer to buy planting material every year, and may represent a substantial cost.

In response to these dangers, genetic engineers argue that, as only a few genes are transferred with these techniques, representing a fraction of the total number of genes present, the dangers are limited. This may indeed be the case with some crops, such as maize, which depend heavily upon humankind for their survival. However, this point ignores the fact that selection operates on organism structure, function and behaviour (the phenotype) not directly on the genes. If a single gene codes for a character that makes the individuals more competitive than others trying to occupy the same niche, they will survive and multiply in the environment. Indeed, if the transferred genes have little or no effect on phenotype, why bother transferring them in the first place? This is not to say that every gene transfer will increase the competitiveness of the recipient, some clearly would have little effect, whereas others may make the recipient less competitive, compared to the "wild type". It is the characteristics that the introduced genes code for that are important in this respect, not the number of genes *per se*.

In addition, it has been argued that the environmental question marks over genetic engineering are less of a concern in developing countries. The argument, typically presented as naked pragmatism, suggests that people in these countries are more concerned with survival than environmental protection, especially when the environmental threat from genetic engineering is uncertain (Ratledge 1992). Unfortunately, this argument separates survival from the environment, a division commonly applied but which is artificial in the extreme. Environmental degradation to exploit finite or renewable natural resources does occur in developing countries for a whole host of complex reasons, including social and economic pressures (Blaikie 1985, Kates & Haarmann 1992). However, the shift to unsustainable resource use is recognized by the people concerned as a serious long-term problem that needs to be addressed.

If the potential environmental problems highlighted above prove to be real, they will result in as much, if not more, long-term damage in developing countries than in the developed world, and therefore need to be seriously considered and not ignored. Developed countries have the resources to try to overcome any environmental problems that result from the release of genetically engineered organisms. The developing countries do not have this advantage. This is not just a moral point, but is based on the same type of naked pragmatism employed in the argument.

The biodiversity question is also a key one (Frankel 1970, Wilkes 1983). It follows that loss of genes from the gene pool restricts options for overcoming future limitations (new pests, diseases, environmental constraints, etc.). Although now commonly quoted as a disadvantage of genetic engineering, the danger is the same, albeit potentially of a different magnitude, for varieties produced through conventional breeding techniques (Fowler & Mooney 1990, Vellve 1992, Holden et al. 1993). However, to some extent this disadvantage is not as severe as some of the others described above (Farrington & Greeley 1989). To begin with, the dangers are well known and understood, with classic examples now a part of many curricula in universities. Secondly, genes can be stored for many years in specialized facilities and brought into play when needed, albeit with a timelag for incorporation. The International Board for Plant Genetic Resources (IBPGR) was created in 1974 with a specific mandate to co-ordinate the conservation of the world crop genetic resource base (various papers in Brown et al. 1989). However, the storage of genes in such facilities does represent something of a risk as mismanagement or accidents can lead to the loss of much material (Hobbelink 1991). Duplication of material in several different stores would help to reduce the risks of gene loss. For example, current research in this field aims to identify some biotypes that represent the range of genetic diversity present in the gene pool of a species. These "key" biotypes would be duplicated in several different stores worldwide. Also, in the long term it is conceivable that the chemical structure of each gene in the gene pool could be determined and stored in a database. Even if a gene is lost completely, it could be reconstituted and inserted back into the gene pool.

However, it should be said that some of the biotechnologies are relatively benign from an environmental, social or ethical point of view. For example, the use of DNA probes as an aid in conventional plant breeding programs in itself does not have any detrimental effects on the environment. Similarly, diagnostic kits (based on antibody techniques) designed to detect trace amounts of pesticide residue or the presence of a natural toxin in food would surely be a net benefit in any situation. The only serious concern here may rest in the use of animals to produce the antibodies for the assay.

The preceding discussion has focused on the ethical and environmental debate commonly included in discussions of biotechnology. From a development perspective, there are two further issues that are relevant and important; the cost and the promise of biotechnology. Both of these are related, but will be discussed separately in the next two sections.

The cost of biotechnology: research and resources

Although the list of techniques included under biotechnology is large and diverse, many do at least have two things in common: they require a great deal of expertise and technical facilities to develop, and hence are expensive (Macer & Bartle 1990, Brenner 1991b). As these essential prerequisites are predominantly located in developed countries, these are potentially set to gain the most from the application of biotechnology (Greeley & Farrington 1989). In addition, as has been pointed out by others, the kinds of skills required to develop genetic engineering and its related technologies are very different from those required for conventional plant breeding (Busch et al. 1991). Biotechnology is rooted in the relatively new subjects of molecular biology and biochemistry, whereas plant breeding has more in common with the older subjects of agronomy and plant physiology.

Even if the different skills needed for biotechnology are present, little can be done if facilities are in short supply (Brenner 1991a, Thomas 1993). Indeed, given these limits, it is very unlikely that genetic engineering will ever replace conventional plant breeding, even in developed countries. It is more likely that the technologies will form a spectrum, with various combinations of genetic engineering and conventional breeding being employed (Cohen 1990, Brenner 1991a).

The cost of biotechnology in terms of money, facilities, expertise and time, may well result in re-allocation rather than new allocation of limited resources for research. If the products generated are more "useful" than those developed with traditional techniques, then all to the good. If not, precious resources could be poured into generating products of less value than those developed by traditional techniques. Even worse, a bandwagon may have been created that will be difficult to stop or even control. If a country spends much money training biotechnologists, it may then have to continue to spend more and more money to keep them. When resources are scarce, there is an even finer line between wise and poor investments, and planners should be very clear where this line lies. Unfortunately, the glamour and prestige typically associated with biotechnology can cloud judgements. More and more developing countries may well strive to establish an indigenous research base in biotechnology, perhaps as a response to fears about reliance on other countries.

A further issue often overlooked in calculating the cost of biotechnology is the need for proper regulation and monitoring. Products resulting from biotechnology need to be tested for safety and environmental impact, and safeguards need to be carefully enforced. Useful products may also need to be patented. All of these activities require skilled personnel and facilities, and provide a further drain on resources. The types of skills required are also quite different from those required for biotechnology. Unfortunately, these may well prove to be the first casualties in the rush for the biotechnology bandwagon.

The promise of biotechnology: another technical fix?

The costs of biotechnology to a developing country may be acceptable if it can deliver upon its promises. However, even if biotechnology can generate the products, will it really improve the lot of poor people in developing countries?

As with the breeding of new plants and animals, the idea of a technical fix to solve major problems in agriculture at a stroke is also not new. Fertilizers, pesticides, machinery, irrigation schemes, new crop varieties, and so on, have all been heralded at one time or another to be "the final solution" to what are perceived by outsiders to be limiting factors in agriculture. Each has had its successes and failures, and all have contributed in some way to the rapid increase in world food production over the past few decades, but they have usually been pushed as solutions to what is regarded as essentially a problem of production. Stated simply, farmers in developing countries are typically poor and, as they derive their income from agriculture, why not provide the means for them to produce more? This is an apparently logical answer to a straightforward question. After all, are not these the very technologies that are the cornerstones of the "efficient" and "productive" agricultural systems of developed countries?

Unfortunately, or perhaps fortunately, things are not so simple. To begin with, underproduction in itself may not be the major problem. Indeed, maybe the agricultural systems are producing at an optimum level for the environmental and social constraints prevalent in the area, but the farmers are not receiving enough revenue per unit of produce. Maybe the farmers do not even regard field production as a major "problem" to be addressed, preferring instead to improve their crop storage, food processing, health care or transport facilities. In short, they may well have another agenda very different from that of the development worker.

People are complex, as are the societies within which they interact, so perhaps it is the height of naïveté to think that the occasional technical fix supplied by a research station some thousands of miles away will provide all the answers. These fundamental lessons apply to biotechnology as they do to the more mundane technologies of fertilizers and machinery.

Along with the above, it could be argued that, even if technical fixes can provide some benefits for the people, they can only scratch the surface of underdevelopment. Much more important are macroeconomic factors that arise from the essentially competitive nature of the world market and, unless these are tackled, then underdevelopment will continue. Even worse, although technical fixes may be regarded as playing a role, they are also sometimes viewed as a distraction from attempts to tackle the major, and much more intractable, causes of underdevelopment and poverty.

Therefore, as much as we may wish the contrary, it is extremely unlikely that biotechnology will remove poverty and underdevelopment at a stroke; the roots of these problems are far too deeply embedded.

Biotechnology: the players

In developed countries the bulk of the biotechnology research and exploitation is in the private sector and the national and regional public sectors. In developing countries, there is little or no indigenous private sector involvement in biotechnology, although of course it could come from non-indigenous companies. Similarly, the costs and expertise involved severely restrict the involvement of the national research facility in many of the developing countries.

The involvement of private enterprise in biotechnology, especially that of the transnational companies, has provoked a great deal of debate among writers on the topic (Walgate 1990, Busch et al. 1991, Hobbelink 1991). As these companies are in biotechnology for a profit, and are typically resource-rich, they do represent an easy target. For example, with genetic engineering, the drive for profit has inevitably produced clear market strategies. To begin with, there has been much emphasis on the lucrative pharmaceutical and food additive markets, where there are potential windfalls to be made. In agriculture, there have been two clear approaches to date. First, there is a focus on a narrow range of crops (wheat, maize, rice, tomatoes, potatoes, etc.) that are grown by farmers who can afford to pay for the genetically engineered varieties. Crops that are important in developing countries, such as cassava, yam, sorghum and millet, have received little if any attention from the private sector. Secondly, some companies have attempted to link the sale of genetically engineered crop varieties with the sale of another product by the same company, usually a herbicide. Such a linkage does not help their position in terms of public sympathy. Pesticides always represent an emotive issue, and an argument that defends the production of herbicide-resistant crops on the grounds of economics or a reduction in the use of other, more environmentally damaging, herbicide often falls on unsympathetic ears.

The criticisms of biotechnology companies, although perhaps correct in detail, are largely unfair. After all, one could point out that many other technologies widely employed in agriculture and medicine are almost entirely dominated by the private sector. Why should the domination of private enterprise in biotechnology be regarded with surprise or indignation? Also, the companies make no secret of the fact that their involvement in biotechnology is based on profit. Why should altruism towards developing countries be assumed or expected in their case?

There are three other major players in the development and application of biotechnology: governments, pressure groups and the public. Governments have a major role to play in legislation, but are strongly influenced – especially in developed countries – by the biotechnology industry and pressure groups. The latter have been very vocal and have succeeded in bringing some of the potential environmental problems of genetic engineering to the attention of the public. Indeed, some recent surveys have suggested that the public places more trust in comments on genetic engineering from such pressure groups than its does in comments from the government (Tait 1992). Whatever the rights or wrongs involved, it certainly seems as if the pressure groups have achieved more success in placing their views before the public, and some have pledged to continue if not intensify their campaign (Tait 1992).

146

Whether the influence of pressure groups will continue or wane in the coming years is difficult to predict. There may be a gradual acceptance by the public of genetically engineered products, coupled with a consequent reduction in legislative restraints. It is also difficult to predict what influence these pressure groups will have upon the governments and peoples of developing countries, but a gradual acceptance of genetic engineering in developed countries will certainly weaken their position. The biotechnology industries, supported by their respective governments, are able after all to offer substantial benefits to developing countries, albeit at a price that may or may not be favourable or clear. On the other hand, the pressure groups may be in the position of always having to stress the potential negatives.

Biotechnology: the competitors

The "players" in biotechnology discussed above represent broad groups. Although these groups interact, each does not necessarily represent a unified body. For example, the companies grouped under the heading "private enterprise" compete with each other in a worldwide market. Countries and trading blocks also compete, reflecting everyone's desire to achieve a good standard of living. What are the implications for biotechnology and the development of these complex interrelationships between the agriculture's of developed and developing countries that form the agrisphere, the agricultural equivalent of the biosphere?

An obvious implication derives from the better resources developed countries have in terms of finance, facilities and expertise, and the consequent competitive advantage in the world economy. Biotechnology can largely be about making money, as one prominent author has recently stated (Ratledge 1992). In addition, the greater potential to develop and exploit biotechnology places developed countries at an adaptive advantage when faced with a changing global environment.

All kinds of scenarios are possible here, including the replacement of products imported into developed countries from developing ones with engineered alternatives. Some go as far as suggesting that the resulting high unemployment, poverty and misery generated in the developing countries could lead to all sorts of doomsday scenarios, culminating in the extreme of a North–South war (Busch et al. 1989). As always, great care has to be taken in avoiding oversimplification, although a widening of the gap between developed and developing countries is certainly possible, if not probable. Ironically, given the gap between the agricultural productivities of developing and developed countries, some have argued that the former have proportionally more to gain than the latter from the application of biotechnology.

Developed and developing countries exist in a complex and dynamic world economy, and have done for many centuries. Technologies come and go, markets rise and fall, and countries adapt to changing socio-economic environments, albeit some more successfully than others. Economic power can shift around the globe many times within a century, and today's developing country can be tomorrow's economic giant.

Biotechnology does offer a competitive advantage to developed countries in the short term, and although it would be naïve to suggest that such countries would readily give this up entirely, there is no reason to believe that it will not be partly shared with developing countries, in much the same way as other technologies have been. Indeed, how this should be done has been the subject of much interest. Two distinct approaches have become apparent. The first emphasizes the transfer of "biotechnological resources" (equipment, techniques, training, etc.), whereas the second, in contrast, places the emphasis on the transfer of products derived with biotechnological resources in developed countries.

Although popular, the "technology transfer" approach has been criticized for being technology driven rather than problem driven. Critics have also likened it to re-inventing the wheel, and argue instead that institutions in developed countries could be paid to develop products specifically for developing countries (Thomas 1993). From a cynical perspective, one can imagine the latter course of action appealing to governments of developed countries, as it allows them to maintain control of the technologies and to spend their aid money within their borders. In practice it is probable that a combination of technology and product transfer will occur, with different donor and recipient countries showing preference for one or the other.

The world economy has been and still is essentially based on competition within and between countries, a fact that is coming more sharply into focus with GATT, the 1992 Rio summit, the emergence of the Pacific Rim economies, the move towards market economies in the former Soviet block along with the painful experience of each recession. Within this system, developed countries clearly want to win, while lending a hand to others. Given an association between biotechnology and wealth generation, it is difficult to perceive of an electorate in the North voting for a severe drop in living standards as a result of increased competition from the South catalyzed in part by biotechnologies developed in the North. The growing spectre of economic defeat would be a very daunting prospect.

Genes: a saleable resource?

There are other issues involved in the incorporation of biotechnology within the agrisphere, besides competitive advantage attributable simply to utilization. The one most commonly considered is the ownership of genes. In marked contrast to the idea of common ownership and availability of genetic resources (Frankel 1974, Wilkes 1983), it has been argued that developing countries have sole access and rights over their genetic material, and this could be of great value in terms of information, rather than as a physical resource (Walgate 1990).

Although historically it can be shown that traditional plant breeding programmes in developed and developing countries have benefited from the "common ownership" precept (Kloppenburg 1988, Juma 1989), there are problems associated with utilizing exotic genes. Unfortunately, the premise supposes that the useful genes can first be identified among the billions present in a wild plant population, that "ownership" can be clearly assigned, and that the country can effectively sell the resource to an interested party. This may at first sight appear to be a relatively straightforward process, but naturally that party will want recompense for the cost of research and development needed to find and use the gene. In addition, only humans recognize geopolitical boundaries and, if the useful gene or group of genes occurs in plants from several countries, there is the danger that they will receive a poor price for their shared resource (Juma 1989). Genes from wild plant species are a resource that, for the moment at least, requires expertise from developed countries for exploitation, and they can and probably will extract a heavy price for their involvement.

No doubt some developing countries or even individuals will forge one-off deals with biotechnology companies, but the companies may have access to powerful economic and political resources. In a competitive market, these resources will be brought to bear to extract the best deal possible for the company, not necessarily for the developing country. Unless an arrangement can be implemented that places a predetermined floor on the level of recompense, the market will find the level itself.

In addition, it must be remembered that many institutions, public and private, within developed and developing countries already hold substantial stores of plant material (germplasms) collected over many years from virtually every corner of the globe. Some of the larger collections in developing countries have been paid for by developed countries, and it is doubtful whether the latter would agree to relinquish their current rights of free access to the material. Therefore, developed countries already have access to a wide range of genetic material that has only partly been characterized, let alone exploited, and may well choose to use this more fully, rather than enter into expensive agreements with developing countries. Indeed, there would also be scope to produce new genes by causing mutations in existing genes from the collections.

Even worse than the above, from the point of view of developing countries, it is also conceivable that in the not too distant future "molecular engineers" could design a useful product and construct a gene to produce it. If the ownership of genetic information is strongly defined at the insistence of developing countries, they could find themselves unable to pay for such gene constructs.

Individual genes, therefore, may not be such a saleable resource as would be imagined. Deals may be forged, but the biotechnology industries and institutions of developed countries are in a strong position. They have the capital for investment, and can look elsewhere for genetic information.

Biotechnology: can it provide appropriate solutions?

Applying biotechnology to help resource-poor farmers in developing countries is in many ways no different from applying other technologies. As always, the emphasis should be placed very much on working with farmers and allowing them to determine what they perceive as the major difficulties that need to be faced (Bunders & Broerse 1991). All to often, research agendas are driven by what planners and scientists perceive to be the problems, with the result that much time and money are spent in producing inappropriate solutions (Chambers et al. 1989). This fundamental error continues to this day, although strong efforts are being made by many to address it (Bunders & Broerse 1991). The development literature is now replete with the words inter- and multi-disciplinarity, but fences erected between disciplines, usually by their practitioners, are often hard to scale.

Biotechnology is no different, although the potentially greater drain on limited resources does mean that failure to address "real" problems has a more detrimental impact on the system as a whole. Indeed, the high tech and glamorous image of biotechnology may even lure people away from considering more traditional and "low tech" alternatives (Hobbelink 1991) that may be more beneficial and certainly much cheaper.

It could also be argued that, if the biotechnologists are more removed from farmers than the plant breeders and agronomists, the probability of producing inappropriate solutions is greater. The emphasis may well become one of "transferring" technologies and products between scientists and institutes of different countries, rather than one of genuinely trying to solve problems. The danger becomes greater when committees and panels that review and allocate funding for agricultural research in developing countries become dominated by biotechnologists, with relatively little if any input from those more able to discover the priorities and needs of the farmers that the biotechnology is ultimately intended to serve.

Some conclusions

The agrisphere is old, complex and dynamic, and biotechnology is but one new component of this system. The nature of biotechnology certainly gives the edge in this system to those countries that have the expertise, structures and money to exploit it, and the results of this advantage may be detrimental to some developing countries. On the other hand, there is no reason to suppose that some techniques or products generated with the techniques will not be shared with developing countries in much the same way as others have been.

The agrisphere has been evolving for many centuries, and the nature of the components and their interactions have changed many times. Will biotechnology on its own really shift the system into instability and collapse, when many other major changes and advantages, including strong protectionism, for the developed countries have not? It seems unlikely, especially with GATT opening the closed markets of some developed countries. The rich countries of the North are not rich, nor the poor countries of the South poor, just because of biotechnology.

Just as it seems unlikely that biotechnology in itself will collapse the agrisphere, it is also unlikely that it will solve all the resilient inequalities within humankind. The causes of these are simply far to complex and intractable to be solved overnight. Similarly, the rich genetic resource base held by some – not all – developing countries may well prove to be of marginal benefit.

However, although biotechnology may not solve underdevelopment at a stroke, its application could help the people of developing countries to improve their absolute rather than relative standard of living. Apart from the major moral and environmental questions that biotechnology raises, the way in which it is applied is of critical importance. Unfortunately, here again there may be more cause for concern. It has not proved to be easy to adapt research in the traditional technologies of plant breeding and agronomy to a more farmer-orientated approach (Chambers et al. 1989), but at least the practitioners are used to dealing with whole plant and cropping systems, and many, at least in the USA, have farming backgrounds (Busch et al. 1991). Biotechnology, on the other hand, is firmly rooted in cellular and molecular biology, and its practitioners have little involvement with whole plants and even less with cropping systems.

The danger is that in the well meaning drive to use biotechnology for the benefit of small resource-poor farmers in developing countries, the emphasis will be on the technologies, products and politics, and not on the real problems faced by farmers (Bunders & Broerse 1991, Ratledge 1992). Technology and product transfer could become the new bandwagons, rather than problem solving. In itself this may not matter, but scarce resources already under pressure could be diverted into this process.

The answer to this potential danger is fundamentally no different from that already in place and gaining ground, and discussions of the farmer-orientated approach can be found in many publications, for example Chambers et al. (1989). Whether these ideas can be brought to bear on the application of biotechnology remains to be seen, but realization may take time. Biotechnology certainly has much to offer to all, but its use in developing countries needs to be carefully targeted to provide true benefits and not be a drain on scarce resources.

It is unrealistic to expect private industry to play a major altruistic role funding the utilization of biotechnology for the benefit of developing countries. The emphasis has to rest firmly with national governments and the international research networks and aid agencies. Private industry may well have a role to play and indeed be more than willing to become involved, but this will usually have to be paid for in one way or another. Indeed, the association of biotechnology with wealth generation may provide some very awkward questions for governments and institutions of developed countries. How far will they be prepared to go in helping developing countries use the new biotechnologies?

References

Anon 1991. *Herbicide tolerant plants: weed control with the environment in mind*. Haslemere: ICI Seeds.

Anthony, K. R. M., B. F. Johnston, W. O. Jones, V. C. Uchendu 1979. *Agricultural change in tropical Africa*. Ithaca, New York: Cornell University Press.

Arus, P. & J. Moreno-Gonzalez 1993. Marker-assisted selection. In *Plant breeding: principles and prospects*, M. D. Hayward, N. O. Bosemark, I. Romagosa (eds), 314–28. London: Chapman & Hall.

Blaikie, P. 1985. *The political economy of soil erosion in developing countries*. Harlow: Longman.

REFERENCES

Boethel, D. J. & R. D. Eikenbary (eds) 1986. *Interactions of plant resistance and parasitoids and predators of insects.* Chichester: Ellis Horwood.

Brenner, C. 1991a. *Biotechnology and developing country agriculture: the case of maize.* Paris: OECD.

— 1991b. Biotechnology in the developing world. *OECD Observer* 171, 9–12.

British Medical Association 1992. *Our genetic future.* Oxford: Oxford University Press.

Brown, A. H. D., O. H. Frankel, D. R. Marshall, J. T. Williams (eds) 1989. *The use of plant genetic resources.* Cambridge: Cambridge University Press.

Bunders, J. F. G. & E. W. Broerse (eds) 1991. *Appropriate biotechnology in small-scale agriculture: how to reorient research and development.* Wallingford: CAB International.

Busch, L., W. B. Lacy, J. Burkhardt, L. R. Lacy 1991. *Plants, power and profit.* Cambridge, Mass.: Basil Blackwell.

Chambers, R., A. Pacey, L. A. Thrupp (eds) 1989. *Farmer first.* London: Intermediate Technology.

Cohen, J. I. 1990. International donor support for agricultural biotechnology. *Food Policy* 15, 57–67.

Cromwell, E. & S. Wiggins 1993. *Sowing beyond the state.* London: Overseas Development Institute.

Dawkins, R. 1986. *The blind watchmaker.* Harlow: Longman.

van Emden, H. F. 1987. Cultural methods: the plant. In *Integrated pest management*, A. J. Burn, T. H. Coaker, P. C. Jepson (eds), 27–68. London: Academic Press.

Farrington, J. & M. Greeley 1989. The issues. In *Agricultural biotechnology: prospects for the Third World*, J. Farrington (ed.), 7–26. London: Overseas Development Institute.

Fincham, J. R. S. & J. R. Ravetz 1991. *Genetically engineered organisms: benefits and risks.* Buckingham: Open University Press.

Fowler, C. & P. Mooney 1990. *The threatened gene.* Cambridge: Lutterworth Press.

Fox, M. 1990. Transgenic animals: ethical and animal welfare concerns. In *The Bio Revolution: cornucopia or Pandora's box*, P. Wheale & R. McNally (eds), 31–45. London: Pluto.

Frankel, O. H. 1970. Genetic dangers in the Green Revolution. *World Agriculture* 19, 9–13.

— 1974. Genetic conservation: our evolutionary responsibility. *Genetics* 78, 53–65.

Greeley, M. & J. Farington 1989. Potential implications of agricultural biotechnology for the Third World. In *Agricultural biotechnology: prospects for the Third World*, J. Farrington (ed.), 49–65. London: Overseas Development Institute.

Gros, F. 1989. *The gene civilization.* New York: McGraw-Hill.

Hobbelink, H. 1991. *Biotechnology and the future of world agriculture.* London: Zed.

Holden, J., J. Peacock, T. Williams 1993. *Genes, crops and the environment.* Cambridge: Cambridge University Press.

Juma, C. 1989. *The gene hunters: biotechnology and the scramble for seeds.* London: Zed.

Kates, R. W. & V. Haarmann 1992. Where the poor live. Are the assumptions correct? *Environment* 34, 5–29.

Kloppenburg, J. R. 1988. *First the seed: the political economy of plant biotechnology.*

Cambridge: Cambridge University Press.

Krimsky, S. 1985. *Genetic alchemy*. Cambridge, Mass.: MIT Press.

Linzey, A. 1990. Human and animal slavery: a theological critique of genetic engineering. In *The Bio Revolution: cornucopia or Pandora's box*, P. Wheale & R. McNally (eds), 175–188. London: Pluto.

Lipton, M. 1989. *New seeds and poor people*. London: Unwin Hyman.

Macer, R. & I. Bartle 1990. *Crop biotechnology in the developing world*. Haslemere: ICI Seeds.

Morse, S. & N. McNamara 1994. *Evolutionary on-farm research*. Discussion Paper, School of Development Studies, University of East Anglia.

Pepper, D. 1984. *The roots of modern environmentalism*. London: Routledge.

Persley, G. J. 1990. *Beyond Mendel's Garden: biotechnology in the service of world agriculture*. Oxford: CAB International.

Ratledge, C. 1992. Biotechnology: the socio-economic revolution? A synoptic view of the world status of biotechnology. In *Biotechnology: economic and social aspects*, E. J. DaSilva, C. Ratledge, A. Sasson (eds), 1–22. Cambridge: Cambridge University Press.

Reijntjes, C., B. Haverkort, A. Waters-Bayer 1992. *Farming for the future*. London: Macmillan.

Richards, P. 1985. *Indigenous agricultural revolution*. London: Hutchinson.

Rogers, E. L. 1990. The human genome project. In *The Bio Revolution: cornucopia or Pandora's box*, P. Wheale & R. McNally (eds), 220–23. London: Pluto.

Ryder, R. 1990. Pigs will fly. In *The Bio Revolution: cornucopia or Pandora's box*, P. Wheale & R. McNally (eds), 189–94. London: Pluto.

Straughan, R. 1992. *Ethics, morality and crop biotechnology*. Haslemere: ICI Seeds.

Tait, J. 1992. Who's afraid of biotechnology? *New Scientist* 134, 49.

Thomas, S. M. 1993. *Global perspectives 2010: the case of biotechnology*. FAST Programme Report, Commission of the European Communities, Brussels.

Vellve, R. 1992. *Saving the seed*. London: Earthscan.

Walgate, R. 1990. *Miracle or menace? Biotechnology and the Third World*. London: Panos Institute.

Wilkes, G. 1983. Current status of crop germplasm. *Critical Reviews in Plant Sciences* 1, 133–81.

CHAPTER SIX

Coexisting with nature in a developing world

Michael Stocking, Scott Perkin, Katrina Brown

Editors' introduction

The need to combine conservation and sustainable utilization of biodiversity was one of the central themes of the Earth Summit at Rio (June 1992). The Convention on Biodiversity attracted substantial publicity, but in fact was initially conceived a decade before the Earth Summit. The germ of the idea was planted by the World Conservation Union in the 1970s and later promoted by the United Nations Environmental Programme (UNEP). The central theme of the Convention is a statement of the importance of biological resources and the need for their sustainable use, rather than just conservation. Interestingly, emphasis is placed both on the ownership of biological resources and the indigenous knowledge that may help their utilization. The Convention states the:

> desirability of sharing equitably benefits arising from the use of traditional knowledge, innovations and practices relevant to the conservation of biological diversity and the sustainable use of its components
>
> Convention on Biodiversity

The question of ownership and exploitation of biological resources provided one of the major difficulties at the Summit. In contrast to the historical view of "global ownership" of biological resources such as genes, Rio "reaffirmed that States have sovereign rights over their own biological resources" (Convention on Biodiversity). Just how these rights were to be put into practice was not tackled at Rio except as broad calls for the respecting of "intellectual property rights" and marketing arrangements that should be "fair and favourable".

The Rio conference identified the main causes of biodiversity loss as being anthropocentric in nature. However, it also stressed the critical role played by biological resources in human welfare and development. If the value of these resources was recognized, and perhaps realized through increased commercialization in some cases, there may be more incentive for their conservation and wise management.

In this chapter Mike Stocking, Scott Perkin and Kate Brown summarize the interaction between contrasting needs for conservation and utilization of biological resources. Echoes of the same debate can be found, particularly in the chapters on biotechnology (Ch. 5) and sustainability (Ch. 2). The authors also discuss two case studies from Africa that illustrate the practical approaches that have been implemented to try and reconcile conservation of biological resources with their use by local people.

Poachers killed

Zimbabwean game wardens have killed all six members of a rhino-poaching gang from Zambia in the last ten days. Wardens have killed 34 poachers and captured 14 in battles this year. Poachers killed 27 of the rare black rhinos and one square-lipped rhino in the same period. *Eastern Daily Press*, 10 October 1990

Introduction: perceptions of a conflict

The needs of local people for economic development are in many developing countries apparently competing with the desires of conservationists to protect endangered species. In Zimbabwe, as the quotation above demonstrates, the conflict sometimes results in death. This chapter analyzes this perceived conflict. It reviews the need for biodiversity conservation, and past and current policies and approaches to conservation.

Historically, conservation policy has concentrated on protecting species and their habitats by the exclusion of people (exceptions being, of course, tourists and scientists). This prohibition tends to identify local resource users as the primary causal agents in loss of biodiversity. Such policies have, therefore, been conflictual and not often successful; local people have paid high costs in terms of loss of access to resources and development opportunities foregone, and relations between conservationists and land users have usually been antagonistic. There is now a growing consensus, reflected in both conservation and development discourses, that if conservation is to succeed and if the needs of local people are to be met, then a different approach is required. This chapter questions whether there can be a "middle road" where both the needs of society and of conservation are met, where species can be conserved within agriculturally based communities ("agrodiversity") and where the mutualities of interest between development and biodiversity can be exploited to the ultimate benefit of all – the "win-win" strategy espoused by the World Bank (1992). Trends in international thinking about the role of conservation and care for the Earth will be described to show how priorities have evolved, and there are initiatives among the development agencies to recognize that nature and society can coexist. This chapter reviews the experiences of two attempts at linking conservation and development; the Ngorongoro Conservation and Development Project, and the East Usambaras Agricultural Development and Environmental Conservation Project, both of which are in Tanzania. Why protect biodiversity?

What is biodiversity?

Biodiversity or, more strictly, biological diversity, refers to the variety, or the number, frequency and variability of living organisms and the ecosystems in which they occur. Biodiversity therefore encompasses species diversity, which is the number and frequency of species of plants, animals and microorganisms; genetic diversity, being the variety of genes within each species; and ecosystem diversity, the different types of plant–animal assemblages and their associated ecological processes.

Decline in biodiversity includes, therefore, all those changes that have to do with reducing or simplifying biological heterogeneity, from individuals to regions. Much of our knowledge about biodiversity focuses on the species level. This is because, at least until recently, species were the most easily identified and measured level of biological organization. Biodiversity is often mistakenly assumed to be equivalent to species diversity and, as a result, much conservation policy is focused on maximizing species diversity. Other considerations are whether a species is restricted to a particular area of habitat ("endemism"), and its degree of rarity or threat of extinction.

We still do not know the total number of species on Earth; estimates range from 5 million to 30 million. Wilson & Peters (1988) have made an assessment of the current status of biological diversity: they list a total of 1 392 485 known organisms, each of which is a repository of an immense amount of genetic information. Each species contains a wealth of genes, from about 1000 in bacteria to 400 000 or more in flowering plants and animals. Then each species contains many individuals: for example the 10 000 ant species comprise about 10^{15} living individuals at any one time. Each individual is genetically unique (except in the case of parthogenesis (the development of a new individual from an unfertilized egg) and identical twinning). The extinction of only one species consequently reduces genetic diversity in absolute terms by a colossal amount. Any loss in genetic potential is a total loss for humankind that cannot be retrieved: this is the essence of the conservationists' case for the protection of biodiversity. Later sections of this chapter will explore the different arguments for the conservation of biodiversity, but first we need to establish whether diversity is being lost, and at what rate, and which are the main causes.

Extinction rates

No precise estimate of the number of species being lost can be made because, quite simply, we do not know the number that exists now. The vast majority of species are not monitored. However, there is no doubt that extinction is proceeding faster than it did before 1800. It seems likely that major episodes of species extinction have occurred throughout the past 250 million years at regular intervals of approximately 26 million years. According to Wilson (1988) the current reduction of diversity seems likely to approach that of the mass extinctions of the Palaeozoic and Mesozoic eras, the most extreme in the past 65 million years.

Species are lost for a variety of reasons. Habitat loss and degradation are the most important causes of the present extinction crisis, but over-harvesting, the introduction of exotic species, and pollution also contribute. Traditionally, from the Darwinian perspective, extinction is the fate of species that lose in the struggle for survival. Taken to its logical conclusion, this view implies that extinction is a constructive process, eliminating obsolete species. However, it is now widely recognized that this is not the case; many extinctions are destructive, and a species' ultimate demise is not a reflection on its "goodness" as a biological organism.

Table 6.1 shows some recent estimates of extinction rates. Many of these estimate potential losses of species by the extrapolation of rates of habitat destruction and calculation of associated extinctions using species area curves. This is based on the principles of island biogeography, which recognizes a relationship between the number of species present and the area of a given habitat (MacArthur & Wilson 1967). There are some problems associated with the use of this rather oversimplified equation to calculate rates of extinction, and it seems likely that the figures determined in this way may be *underestimates* of the expected extinction rate.

Unlike the major episodes of extinction in the past, most present losses occur as a result of human action, particularly the loss of natural habitats.

Table 6.1 Estimates of species extinction.

Estimate of species loss	Global loss per decade (%)	Method of estimation	Reference
1 million species 1975–2000	4	Extrapolation of past exponentially increasing trend	Myers (1979)
15–20% of species 1980–2000	8–11	Species-area curves	Lovejoy (1980)
25% of species 1985–2015	9	Loss of half species in area likely to be deforested by 2015	Raven (1988)
2–13% of species 1990–2015	1–5	Species area curves	Reid (1992)

Source: Reid (1992).

The destruction of tropical rainforests has been identified as a particular cause of species extinctions. This is in part attributable to their high biological diversity (see Myers 1984), but also because of the rapid rate of expansion of the agricultural frontier that has occurred in many tropical countries in the past few decades. The *proximate cause* of deforestation is readily identified as people cutting down trees, clearing land for conversion to other uses, but the *fundamental causes*, the driving forces fuelling the destruction of natural habitats and loss of global biodiversity, lie in the workings of national and international political and economic systems (Brown & Pearce 1994). Conservation policies that tackle the proximate causes will not be effective unless the fundamental causes – distorting policies and incentives, inequitable distribution of land and other resources – are also addressed. It is easier to blame the local land user than to examine the underlying political and socio-economic reasons for their actions.

Rationale for conserving biodiversity

Having established that biodiversity is being lost, and apparently at an accelerating rate and primarily as a result of human actions, what is the rationale for its conservation? It is possible to characterize two paradigms that represent the extreme views supporting the conservation of biodiversity (Machlis 1992). The first is utilitarian and it considers a lost species as a lost commodity. Machlis argues that this biotechnological stance redefines natural variation into a "pharmaceutical and industrial warehouse", so that decline in biodiversity represents a potential reduction in future stock value, dividends and profits. It is this view that is conventionally assumed to underscore an economic analysis of biodiversity loss and conservation strategies. At the opposite end of the spectrum is the argument that species have intrinsic value, a central tenet of the Deep Ecology movement. Biodiversity is framed as a moral condition, its preservation a moral responsibility based largely on the rights of non-human species and never subject to the whims of human preferences.

In between these utilitarian and Deep Ecology perspectives lies a range of arguments in support of the conservation of biodiversity, and four broad categories can be identified. The arguments are characterized as economic, ethical, scientific, and aesthetic. We will briefly describe each in turn, although there may be some overlap, as for example, between scientific, aesthetic and economic arguments[1].

Economic arguments

Biodiversity brings many benefits to humans (Table 6.2). These include direct uses by humans for consumption and production, for example in the provision of food, fuel and shelter. The economic value of these is recognized; many products are traded in markets, whereas others may be subsistence products, but critical to the livelihood needs of the majority of the world's population, especially in developing countries. Such resources include wild foods and bush meat, products that normally fall outside conventional economic analysis. In addition to these uses and values, a whole range of other economic values of diverse biological resources are recognized by the growing discipline of *environmental* or *ecological economics*. These involve *indirect use values*, which include ecosystem functions such as watershed protection, carbon and waste cycling, and environmental buffering, as well as non-consumptive uses such as recreation. *Non-use values* incorporate *option value*, or the future uses given a change in knowledge or technology, and *existence values*, the benefit humans get from knowing that species exist. A burgeoning literature now exists that discusses these different values and the various methods for their measurement (Pearce 1993 provides an introduction). Such discussion is beyond the scope of this chapter, suffice to say that there are many different views as to the scope and lim-

---

1. Some environmental economists would argue that economic valuation techniques should be able to account for all scientific (ecological, functional, future or option values) and aesthetic (recreational, existence and bequest values) aspects. The concept of total economic value encompasses all these aspects, each of which confers welfare to human beings; all such values are anthropocentric, instrumentalist and based on human preferences (for a fuller explanation of TEV, see Pearce 1993). This leads to the distinction of two arguments for conservation, the "rights approach" and a "trade off approach" as identified by Pearce (1994).

Table 6.2 A classification of benefits and potential values of biological diversity (adapted after McNeely 1988 and Spellerberg & Hardes 1992).

Direct values	
Consumptive use	Value placed on nature's products that are consumed directly without passing through the market: 1. Source of food for local communities 2. Construction materials (wood, poles etc.) 3. Raw materials for local manufacture and goods 4. Medicines, potions and herbs 5. Fuelwood
Production use	Value on products that are commercially harvested: all of the consumptive use products above which enter the market, plus: Contributions to the production of domesticated resources: 1. Wild species as source of new domesticates 2. Wild genetic resources to improve established domesticates 3. Rangeland and wild forage contribution to livestock production 4. Wild pollinators (essential for many crops) 5. Natural pest control

Indirect values	

Non-consumptive use	Generally, the value of benefits accruing to society in the form of services which are not consumed or traded:
	1. Leisure pursuits such as bird-watching
	2. Aesthetic value of seeing, hearing, touching or being with wildlife and natural ecosystems
	3. Environmental maintenance functions:
	  – regulation of macro- and micro-climate
	  – maintaining CO_2–O_2 balance
	  – storage and cycling of nutrients/water/materials
	  – absorption of pollutants and wastes
	  – soil formation and prevention of erosion
	  – protection of water catchments
	  – regulation of water flow
	4. Environmental change indicators
	5. Protection from natural environmental variability and buffering extreme events: e.g. floods, typhoons, landslips
	6. Photosynthetic fixation of solar energy
	7. Ecosystem functions related to reproduction, such as pollination, gene transfer, cross-fertilization

Option	A value based upon the uncertainty of the future and the need to avert risk. This is the "option" of having future access to a species or ecosystem. Similar to "serendipity value" or the potential that each species may yet yield an undiscovered human use such as food or medicine or specialist function (Pearsall 1984).
Existence	Ethical, moral and aesthetic values; a value attributable to nature by simply being there, unused and unvisited – reflects the sympathy, concern and responsibility that some people feel for species; e.g. whales and pandas.

itations of the economic value of biodiversity (Brown & Moran 1994). However, quantification can help make choices between competing or alternative policies with different environmental impacts. Like any economic decision, investment in conservation is a question of scarcity and choice. Economic analysis may therefore be of use in policy analysis and in highlighting various perverse incentives that encourage destruction of the environment, because they fail to take account of full costs of action (see Repetto & Gillis 1988).

Ethical arguments

There is a moral and ethical argument for conservation of biodiversity based on the inherent rights of species to exist. This stance argues that species have an intrinsic worth, regardless of their actual or potential use to either humans or other species[2], and that one species, *Homo sapiens*, does not have the right to a obliterate another. For many authors this last argument is the most powerful, expressing the duty of stewardship of the natural world. As Ledec & Goodland (1988: 14) put it: "For a couple of generations of human beings to eliminate unnecessarily a sizable proportion of the diversity of life on Earth can be construed as an act of considerable arrogance". One difference between the economic and intrinsic value approaches is that economic values can, in principle at least, be measured. Intrinsic values cannot, so that the ethics argument may be *absolutist* and therefore of limited use in making policy decisions, as it offers no option of trade-offs between competing uses.

Scientific arguments

The role of biodiversity in maintaining the Earth's life-support systems is as yet poorly understood. We still do not know exactly how biological diversity contributes to ecosystem characteristics such as *resistance* and *resilience*, and we cannot define with any certainty which organisms or species are critical to ecosystem functions. The most often used analogy likens the loss of species to the rivets holding an aeroplane together being progressively popped one by one. Although the aeroplane continues to fly, despite the loss of rivet after rivet, at some point too many rivets will have been lost and the plane will fall apart, with disastrous results. We do not know how many "rivets" we can afford to lose, what the impact on the "aeroplane" and its passengers may be, and when the crash will occur.

---

2. For more about these ethical arguments, see Armstrong & Botzler (1993), Rolston (1988) and Naess (1989).

In addition, biodiversity is critical to many fields of scientific enquiry, such as evolutionary biology. Undisturbed natural habitats may be essential if we are to be able to monitor and understand the impact of human activities on the environment. Biodiversity is also necessary for future evolution ("evolutionary value" defined by Juma 1989): if we destroy diversity we are depleting the genetic base for species changes and possible future evolutionary developments.

Aesthetic arguments

Biodiversity brings many intangible benefits to human kind that are deeply rooted in our cultural appreciation of nature and beliefs about creation. These attributes are not easily subsumed in economic valuation (see Brown 1994) and, indeed, many people may find attempts to put economic values on these characteristics morally repugnant. Biodiversity brings pleasure and richness to people's lives, an abundance that may be completely independent of the more tangible benefits to be derived. Aesthetic arguments will also be culturally specific, often bound to religious belief systems and cosmologies. Although these beliefs often do not appear to conform to economic rationality, they may be the most important determinant of management of natural resources. They do, for example, most often form the basis for disputes between indigenous people and developers.

Conservation has been argued on the grounds of immediate economic rationality, on moral and ethical principles, on the support that biodiversity provides to almost every field of human endeavour, and on emotional attachment. These are powerful arguments. But are they supportable in the face of apparent conflicting demands for development and the major ethical dilemma of human life versus nature? The following sections focus on protected areas, the international attempts to link development goals to conservation, and two examples from East Africa to provide empirical case studies.

Policy approaches

The most effective way of protecting biodiversity is to protect natural habitats. Zoos and botanical gardens can hope to conserve only a small proportion of the Earth's biodiversity, and they also suffer from the disadvantage that they remove organisms from interactions with their natural environment and thus prevent many ecological and evolutionary processes. For this reason, over the years a great emphasis has been placed on establishing national parks and nature reserves. However, protected areas have been imposed with scant regard for the needs of local people. People have lost their rights of access to, and use of, productive resources, and the establishment of protected areas has imposed high costs on already impoverished communities, when the main beneficiaries of these policies have been the international community and a few rich tourists (Wells 1992). Hostility and conflicts, as reflected in the quotation at the beginning of this chapter, have very often been the result of heavy-handed conservation policies. However, thinking has now changed and, to an extent, practice has followed. The next section discusses these changes towards a more people-centred approach to the conservation of biodiversity, taking as a starting point the groundbreaking policy document, the *World conservation strategy*.

Conservation and development:
new directions for sustainable biodiversity management

The *World Conservation Strategy* (WCS) was formulated in 1980 by the International Union for Conservation of Nature and Natural Resources (IUCN, now called The World Conservation Union) in collaboration with UNEP, the Worldwide Fund for Nature Conservation (WWF), FAO and UNESCO. Its broad aim was to launch a new message that conservation is not the opposite of development; that humanity is part of, and relies upon, nature, natural resources and ecological processes; that conservation cannot be achieved without development to alleviate the poverty and misery of people. It was a strange mix of pragmatism – farmers cannot conserve for their future unless their immediate survival is also assured; rhetoric – humankind's "life of dignity", for example; and a new intellectual framework – the term "sustainable development" attained its first widespread exposure in the WCS document (IUCN/UNEP/WWF 1980). Yet its provenance was clearly ecological and conservationist.

Three objectives were specified:
- essential ecological processes and life-support systems must be maintained
- genetic diversity must be preserved
- any use of species or ecosystems must be sustainable.

In other words, two non-negotiable conservation objectives are placed before a third objective specifying a natural resource use conditionality of sustainability. The WCS devotes most attention to the linkage between conservation and development and how the aims might be achieved in practice, and in essence states that:

- Exploitative development is wrong. Projects that either ignore the consequences of their actions or invite conflicts between nature and development should not proceed. The implication was that nearly all development projects up to 1980 failed the WCS test.
- There are rational reasons why conservation is in the interest of humanity. Natural resources can have a value ascribed to them, and it is calculated self-interest that should make us want to conserve.
- Development needs attaching to conservation so firmly that the two are effectively contingent on each other.

These were challenging and innovatory statements, and they came at an opportune time. The evidence for species loss was accumulating, as was the rapidly accelerating loss of unprotected habitats such as humid tropical forests. Development practitioners were seeing expansions in aid budgets but with little to show for the expenditure; agricultural technology transfer projects routinely failed for both technical and social reasons; so-called "integrated" rural development projects were little better than their sectoral counterparts; project evaluations usually showed that predictions of economic returns on investment in developing countries were often wildly optimistic. Linking conservation and development threw a lifeline to some and ensured that, arising from the WCS and its successors, some key actions would be incorporated progressively over the next decade.

Successors to the WCS have largely kept the same message. If anything, the conservationist stance has moderated further. For example, *Caring for the Earth* opens with its principal aim "to help improve the condition of the world's people" (IUCN/UNEP/WWF 1991: 3). Over the decade leading to the Earth Summit in Rio de Janeiro in 1992, the main progress was a strengthening of the concept of "sustainable development", stronger articulation of the notion that conservation could be achieved by distributing the benefits of development and the costs of preservation more equitably, and publicizing the economic argument that conservation really does pay.

Caring for the Earth proposed that special measures are needed to safeguard species and genetic stocks, and that areas be designated for their importance in biological diversity. Called Biodiversity Conservation Regions (BCRs), these areas would have individual strategies of protection developed by local communities, government agencies and other interest groups in order to provide for human development in ways that conserve biological diversity. In short, BCRs were seen to be part of a system of diverse protected areas with explicit policies but multiple functions.

In summary, the current guidelines stress: accentuating the positive benefits of conservation, especially to local communities; providing for legitimate development interests, especially for the poor; avoiding wherever possible conflicting interests between conservation and development; and empowering local groups to plan and manage their own resources. But formally promulgated protected areas remain.

Arising from the World Conservation Strategy and its successors are two essential thrusts:

- the limitation of resource use in order to protect biodiversity
- the control of human population increase.

Inherent in such reactions is the simplistic assumption about finite natural resources and the threat posed by population growth. Also inherent is the inevitability of conflict. First, there are the conflicting demands between short-term private and long-term societal interests. Secondly, there are problems as to who pays and who benefits. Thirdly, there is the moral dilemma about which species is to be preserved, *Homo sapiens* or some "natural" organism. Is reconciliation possible? In the context of national parks and official reserves, many ecologists and international institutions now believe that an accommodation is possible between the rights of local people and the need to conserve biodiversity for society in general.

In arguing for "conservation with a human face", ecologist Richard Bell (1987) makes the valid point that Africa's ecological crisis has been grossly exaggerated. Alarm and despondency characterize the writings of many, for instance, Timberlake's (1985) *Africa in crisis* and Pimentel's (1993) review of the status of world soil erosion and conservation paint lurid pictures of billions of tonnes of soil washing to the sea. Such accounts pander to the sensational, and accentuate the prevalence of what must be admitted as extremely degraded but localized zones. Elsewhere, it has been argued that there are rational scientific and career reasons why professionals would wish to make outlandish claims as to the seriousness of soil erosion (Biot et al. 1992, Stocking 1995). Let us, therefore, accept the premise that the conflict may not be quite as ubiquitously serious as some may present it to be, and that the conventional portrayal of degradation and simplistic explanations of causation may be erroneous. An increasing body of work indicates that in some circumstances very different processes may be under way. For example, research by Tiffen et al. (1994) in Machakos District in Kenya refutes the conventional model of increasing human population pressure and demand for agricultural land necessarily leading to environmental degradation. Other studies, such as those by Guha (1989), Peluso (1993) and Broad (1994), reverse the conventional view of poverty and the causes of environmental degradation, and instead show poor people to be activists and defenders of the environment on which their livelihoods depend.

Nevertheless, there are potentially conflicting objectives between people who wish to expand their operational area of activity and conservationists who wish to protect remaining areas of "naturalness". At its most extreme, the parties may be separated by electrified fences and armed game guards, poachers are shot on sight and government staff fear for their safety in local villages. Around the Palamau Tiger Reserve in India, for example, Forestry Department officials have been killed in reaction to restrictions placed on local people's collecting bamboo from the forest. "Hard edges" characterize the boundary between people and parks in many areas. For this reason, there is a growing realization that the long-term protection of biodiversity has to enlist the co-operation and support of local people. As Wells et al. (1992: 2) state, ". . . it is often neither politically feasible nor ethically justifiable to exclude the poor – who have limited access to resources – from parks and reserves without providing them alternative means of livelihood". This is a subset of the more general argument that has blossomed since the late 1980s that sustainable development is achievable only by linking conservation and development through local participation, genuine partnership and empowerment.

Protected areas do differ substantially in their extent, objectives and principal characteristics (Tables 6.3 and 6.4). Some, such as "managed resource protected areas", specifically invite human intervention as part of the management strategy. Others, such as wilderness areas, largely exclude human activity. Increasingly, the major agencies such as the World Bank have new

Table 6.3 Functions and benefits of protected areas: natural, modified or cultivated ecosystems (adapted from IUCN/UNEP/WWF 1991).

Functions of protected areas	Benefits of protected areas
Safeguards for:	Developmental importance through:
1. "Intrinsic, inspirational and recreational values of the world's great natural areas".	1. Conservation of soil and water in vulnerable and sensitive areas.
2. "Life-support systems" in natural and modified ecosystems; conservation of wild species and areas of especially great species diversity.	2. Regulation and purification of water flow (mainly wetland and forest)
3. Culturally important landscapes, historic monuments and other heritage sites.	3. Shielding from natural disasters such as floods and storm surges (e.g. coral reefs; riverine and coastal wetlands; mangroves)
4. Sustainable use of wild resources in modified ecosystems.	4. Maintenance of natural vegetation on sites of low intrinsic productivity
5. Traditional use of ecosystems by indigenous peoples.	5. Maintenance of wild genetic resources and species important to medicine.
6. Recreational and educational use of natural, modified and cultivated ecosystems.	6. Protection of species and populations susceptible and sensitive to disturbance.
7. Scientific research, present and future.	7. Habitat provision for harvested, migratory or threatened species for breeding, feeding or resting.
	8. Income and employment for local people (especially through tourism).

Table 6.4 Protected area categories and management objectives (based on the 1994 revisions to IUCN/UNEP/WWF 1991 discussed at CNPPA meeting prior to XIXth IUCN General Assembly, Buenos Aires).

Type	Management objectives
Categories Ia & Ib: strict nature reserve/ wilderness area	Managed mainly for science or wilderness protection. Objectives: to preserve habitats, ecosystems and species in as undisturbed state as possible; maintain genetic resources and ecological processes; secure examples of the natural environment for science and education. Management is achieved through minimizing disturbance and limiting public access. Wilderness areas may provide for public access and indigenous human communities, and "should offer outstanding opportunities for solitude".
Category II: national park	Managed mainly for ecosystem protection and recreation. Protection of natural and scenic areas for scientific, educational and recreational use. May take into account the needs of indigenous people in so far as no adverse impact on ecological, geomorphological, sacred or aesthetic attributes.
Category III: natural monument	Managed mainly for conservation of specific natural features. Objectives: to protect or preserve significant natural features and maintain their unique characteristics; eliminate exploitation or occupation inimical to conservation; deliver to any resident population such benefits as are consistent with designation.
Category IV: habitat/ species management area	Managed mainly for conservation through management intervention. Objectives: to secure habitat conditions necessary to protect species, biotic communities or physical features where these require specific human manipulation; facilitate scientific research and environmental monitoring; plus objectives as in Category III.
Category V: Protected landscape/ seascape	Managed mainly for conservation and recreation. To maintain the harmonious interaction of people and land while providing opportunities for public recreation and tourism; to support life-styles and recreation which are in harmony with nature.
Category VI: Managed resource protected area	Managed mainly for the sustainable use of natural ecosystems. Objectives: to protect the biological diversity and other natural values in the long term; promote sound management practices for sustainable production; protect the natural resources base for future use; contribute to regional and national development.

policies that recognize the inherent importance of including local people in the planning and benefits of all types of protected areas (Ledec & Goodland 1988).

The following two sections of this chapter evaluate the record of two quite different attempts (but from the same country, Tanzania) to put into practice an explicit linkage between conservation and development. Each case study was an endeavour by a major multilateral environmental organization (IUCN) using bilateral development aid funds (from Norwegian and European Community sources) to demonstrate the workability of World Conservation Strategy aims.

Ngorongoro Conservation Area

Multiple-use areas: linking conservation and development

Of the World Conservation Union's (IUCN 1985) eight categories of protected areas, the last, "Multiple-use management area",[3] has received a high profile as an alternative to traditional national parks. Examples of the range in type of this model of protection are: the Air–Tenere National Nature Reserve in Niger, which supports rare mammals such as the Barbary sheep and addax, dama and dorcas gazelles; the Annapurna Conservation Area in Nepal, with the world's deepest river gorge; and Dumoga–Bone National Park on Sulawesi in Indonesia, established contemporaneously with two irrigation projects on the immediate park boundary. They are all managed for the "sustained production of water, timber, wildlife, pasture, and outdoor recreation, with the conservation of nature primarily oriented to the support of economic activities" (McNeely et al. 1990: 59). Unlike national parks, rather than excluding human consumptive utilization, multiple-use areas seek to control the scale and location of human activities, such that development requirements and conservation objectives can be met within the same protected area or immediately adjacent to the boundary under a unified management. Multiple-use areas thus offer the promise of a more "sensitive" or "human" approach to conservation; in particular, they are a potentially valuable way of promoting conservation in developing countries, without the many conflicts that have often been associated with national parks. As a result, many governments and international agencies are expressing a growing interest in multiple-use areas, and many new areas are being established in Asia, Latin America and Africa.

The Ngorongoro Conservation Area

One early experiment with the multiple-use concept in a developing country was the Ngorongoro Conservation Area (NCA) in northern Tanzania (Fig. 6.1). Established in 1959, the NCA spans some 8300 km², supporting five major land uses:

---

3. At its XIXth General Assembly in June 1994 at Buenos Aires, IUCN accepted a revised categorization of protected areas (see Table 6.4). "Multiple-use management areas" are now subsumed into Category vi: Managed resource protected areas. The notion of "multiple-use" is retained here to highlight the principal theme of this chapter: the explicit linkage of conservation and development objectives within defined areas.

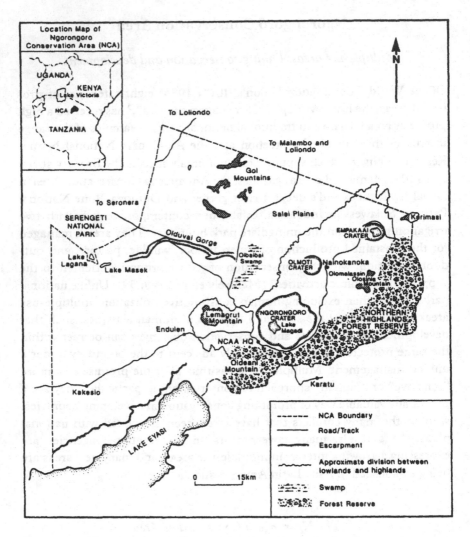

Figure 6.1   The Ngorongoro Conservation Area, Tanzania: location map.

- **Wildlife conservation**   The NCA lies adjacent to the renowned Serengeti National Park, and is a critical component of the Serengeti ecosystem. The short-grass plains of the NCA are utilized each year by the Serengeti's migratory wildebeest, zebra and gazelle, which now number some 2.5 million animals. The NCA also supports a wide variety of resident wildlife, including a small population of the highly endangered black rhino, a diverse population of birds, and several endemic grasses and plants. Ngorongoro Crater – a vast caldera formed by the collapse of an ancient volcano – supports a particularly spectacular assemblage of large mammals.

- **Tourism** Over 100 000 tourists a year visit the NCA, primarily to view wildlife within Ngorongoro Crater. More people visit the NCA than any other tourist destination in Tanzania, making the Area an economic asset of considerable national significance.
- **Pastoralism** Practised in the Ngorongoro region for probably over 7000 years (Bower & Gogan-Porter 1981), pastoralism supports a colourful local population. In 1988, the NCA was inhabited by some 25 000 Maasai pastoralists, and nearly 300 000 head of livestock. In addition to its conservation and tourism responsibilities, the management authority for the NCA has been given a statutory duty to "promote and safeguard the interests" of the Maasai (Government of Tanzania 1975).
- **Archaeology/palaeontology** Some of the world's most important archaeological and palaeontological sites are located within the NCA, including Olduvai Gorge, the site of much of the Leakeys' famous work. Fossils, stone tools, and other artifacts have been recovered from these sites, which have been used to support the theory of human evolution spanning some 4 million years (Mturi 1990).
- **Catchment forestry** The NCA contains several catchment forest areas, the most important of which is the Northern Highlands Forest Reserve. Covering over 800 km^2, this area provides year-round water supplies to the densely settled agricultural communities bordering the NCA, and is a vital regional asset.

Land-use conflicts

This combination of land uses within a single protected area makes the NCA unique in East Africa; it has also made the Area contentious. Since its establishment over three decades ago, the NCA has been characterized by controversy, accusation, and counter-accusation. Hard-line conservationists have portrayed the Area as an ecological disaster in the making, citing human- and livestock-population growth, competition for grazing between wildlife and domestic animals, and growing demands for infrastructure and other socio-economic inputs. Anthropologists, development practitioners and the Maasai themselves have focused on the difficulties resulting from the Area's land-use restrictions (particularly a ban on cultivation, in place from 1975 to 1992), and the lack of material and financial benefits accruing to local communities from the management system. As an example of the politicization of the debate, a Maasai participant at the 1994 Arusha Conference on Pastoralism and the Environment proclaimed that what was being practised in Ngorongoro was worse than the apartheid imposed in South Africa.

Evaluation of the NCA's multiple land-use system

Despite the passion in the arguments, there have been few objective studies of the situation in the NCA, and rhetoric had largely clouded the Area's record of achievement in both conservation and development. In 1987, in response to the mounting concern about the long-term future of the Area, the Government of Tanzania initiated a three-year evaluation of the NCA's multiple land-use system, in collaboration with the IUCN. The evaluation consisted of three principal sets of activities: aerial and ground surveys of the human, wildlife and livestock populations, in order to determine both the distribution of these populations and their trends over time; a series of 14 technical studies on a range of priority conservation and socio-economic issues, involving some 30 national and international specialists; and consultation activities with local communities, the NCA management authority, government officers, and other interested parties.

Both sides in the debate clearly had legitimate concerns, but these were often exaggerated. From the standpoint of conservation, the evaluation concluded that most of the key wildlife habitats had been protected and that most species of wildlife were secure. Despite predictions of an exponentially increasing livestock population, the surveys revealed that numbers of both cattle and small stock were well within previously recorded limits. In many areas, there appeared to be relatively little overlap between livestock and wildlife, and competition for grazing resources was therefore minimal. In the few zones where competition did occur, it was the livestock that suffered at the expense of wildlife.

There were, however, several important conservation concerns. The catchment forest was being degraded, in part through uncontrolled grazing and burning by Maasai pastoralists. On some occasions, fossils within Olduvai Gorge had been trampled by domestic stock, and irreparably damaged. In addition, the human population had increased rapidly, from some 8200 in 1962 to a total of nearly 25 000 in 1988.

From the standpoint of development, a clearer picture of endangerment to the conservation development linkage emerged. Roads, water-supplies, and other infrastructure had fallen into disrepair. Livestock health services were inadequate, and disease was now a major constraint on the size of individual herds and the productivity of the livestock population. The ratio of livestock to people had fallen sharply, and there were signs of poverty and malnourishment among the Maasai. It was striking that there were virtually no mechanisms in place to provide local communities a voice in the management of the Area, or to ensure that they received some proportion of the NCA's sizeable tourist revenues. It was clear that, over time, the management of the NCA had swung progressively away from development concerns, towards conservation objectives as the primary focus.

Despite an apparently bleak evaluation, it is important to place the NCA into context by comparing the socio-economic status of the resident Maasai with that of pastoral groups elsewhere in East Africa. Maro (1990), for example, has suggested that the provision of education and health services within the NCA is roughly on a par with that in other areas of Ngorongoro District. Similarly, McCabe et al. (1992) have shown how pastoral societies throughout East Africa have experienced declines in livestock:humans ratios: the NCA is by no means unique in this regard. In addition, Nestel (1986) has found levels of malnourishment among the Maasai on Kenyan group ranches that are as high or higher than those in the NCA. The implication is that, in this multiple-use area, greater protective effort may have to be accorded to the human population rather than the animals.

On the basis of these findings, the evaluation recommended that the NCA's multiple land-use system should be continued, so long as some important changes were made to ensure that:

- the catchment forests and key archaeological sites were accorded greater protection
- improved planning and zoning systems were established, to control and guide land-use and development activities
- greater emphasis was placed on development activities, particularly those relating to food security and livestock health
- the Maasai became active partners in the management of the Area, and received a share of the NCA's annual revenues.

Multiple-use areas in the developing world: lessons from the NCA

The clearest lesson to emerge from the NCA is that multiple land-use areas do not provide a panacea for integrating conservation and development in the Third World. In particular, the NCA highlights the difficulty of managing multiple-use areas in a balanced fashion, such that both conservation and development objectives are achieved. In the developing world, the lack of resources, the evident human needs, and the incapacity of institutions to manage complex developments, imply that multiple-use areas will be difficult to sustain. The temptation to revert either to development or to conser-

173

vation will be difficult to resist in the face of local political rhetoric or pressure from the environmental lobby.

Nevertheless, multiple-use areas clearly do have a role in the developing world, and in many regions they may well represent the best option for extending some degree of conservation protection to ecologically important areas. Multiple-use areas stand to be particularly useful in populated areas where large tracts of uninhabited land are no longer available for conservation purposes; they also stand to be of use as buffer zones around national parks. However, it will be important for new multiple-use areas to address the concerns raised by the NCA's experience, namely the vital needs for:

- a system of checks and balances, to prevent either conservation or development objectives from gaining undue advantage
- a management system based on the sharing of power and revenue between government and local communities
- participation, which represents the interests of all stakeholders
- a system of land-use zoning and development control that is clear to the local population, viable ecologically and easy to administer
- placing protected areas within a regional development context, such that multiple-use areas are part of an overall strategy for conservation and development, and their existence is not threatened by high rates of immigration, attracted by new development opportunities.

Finally, it is important to stress that multiple-use areas do not replace national parks and similar reserves: they are an additional approach to conservation – not a substitute – and they should be used only where coexistence of nature and economic exploitation can be rationally argued as in the interests of both conservation and development.

Human development in the Usambara Mountains

A managed forest reserve: linking conservation and development

Protection of tropical forests gives the greatest cause for concern among conservationists because of their immense diversity and their vulnerability. Not only is it estimated that tropical forests will account for most of the species extinctions in the near future, but also ecosystem and habitat destruction and global effects such as induced climatic change will have wide-ranging impacts on all humanity. It is widely cited that 20 ha of tropical forest are being destroyed or converted every minute; only 10 per cent of the lost forest is developed as plantation forest; yet 70 per cent of developing country people rely on forests for their fuel and other wood needs.

What is the basis for linking conservation with development of tropical forests? First, many authorities argue that tropical forests are national assets that form a springboard for development. They are a source of employment, a foreign currency earner, the principal source of energy for developing countries, and a reserve of land for future development of agriculture and plantations. In other words, tropical forests should not be frittered away for short-term gain and uncontrolled exploitation. Secondly, an international perspective is relevant to preserving tropical forests; commonly cited are the role of the forests in atmospheric purification, control of global weather and temperatures, cross-border impacts such as floods and sedimentation, and as the prime genetic repository. Thirdly, and most importantly for their protection, tropical forests are locally vital for communities. FAO estimates that 200 million people worldwide live in forests and 2000 million are directly dependent on forest products.

However, forests are vulnerable to destruction. Powerful commercial and political interests are involved in their exploitation; whereas the people subsisting in the forests are usually poor and marginalized. Forest products, especially hardwoods, have substantial value on the world markets, whereas local-use values of forests are relatively meagre from a national perspective. Many persons see forests as the antithesis of development: land title is claimable only if trees are cut down; wicked spirits inhabit jungle; forests signify darkness and illicit activity, but open fields are all-revealing. Such is the stuff of myths. Nevertheless, the arguments to maximize the utilization of trees can be very persuasive to local people, as well as national (and private) exchequers.

The East Usambaras Project Area[4]

"Ecological islands" of submontane tropical forest persist on isolated equatorial mountains in Tanzania. It is only the difficulty of access that has preserved remnants of the forest today. One such "island" is in the Usambara Mountains, which, as part of the chain known as the Eastern Arc Mountains, run from northern to southern Tanzania (Fig. 6.2). Proximity to the coast and an annual rainfall of 2000 mm and more gives rise to some of the most luxurious vegetation in Africa, with trees rising to 65 m and an unrivalled biological diversity (Hamilton & Bensted-Smith 1989). Of 217 tree

---

4. A fuller account of this case study appears in Stocking & Perkin (1992).

species, 50 are endemic or nearly so. Rodgers & Homewood (1982) claim that the East Usambaras are as important for biodiversity as the Galapagos Islands. The biological importance of these mountains is indisputable.

An immediate economic case can also be constructed for the conservation of forests on the Usambaras. Forty thousand people live there, many of whom rely on the forest for fuel, poles and other products. For example, 63 different forest species of plant are used for medicines and 16 for edible fruits. At a regional level, the forests protect the headwaters of the rivers supplying nearby population centres in Tanga Region. The forest also supplies up to 90 per cent of Tanzania's cardamom production, a spice of high value and important to export, but which needs a closed canopy under which to grow. Also, in forest margin areas, much of Tanzania's cinnamon and tea are grown, and smallholders make a substantial living from cloves.

A rationale for linking conservation of the remaining tropical forest areas and the economic development of the local community has been institutionalized in an internationally funded project executed by IUCN, called the East Usambaras Agricultural Development and Environmental Conservation Project, which started in 1987. The case rests on four principal land uses in which improvements initially supported by external funding could bring benefits to forest conservation and to local populations:

- *Subsistence agriculture* Much of the present cultivated land is degraded because of the steep slopes, fragile soil and the type of exploitation. If agricultural practices could be intensified and improved on the existing area, the pressure to clear further areas of forest should diminish.
- *Cash crops* The highest-value crop, cardamom, needs shade. However, research suggests that shade could be provided from within an agroforestry system and the cardamom could be cultivated on permanent sites, instead of depleting the undergrowth in the forest. Permanent cultivation would give farmers greater control over the crop, easier harvesting and more assured income.
- *Fuelwood and pole extraction* Large amounts of wood are removed from the forest. Currently degraded sites could be pressed into productive land use through woodlots and small plantations.
- *Pit-sawing and logging* An improvement in forest management and the establishment of local systems of control by villages should enable a sustainable off-take of timber from the forest. Local communities should benefit more substantially from the operations and would have a stake in preserving future income.

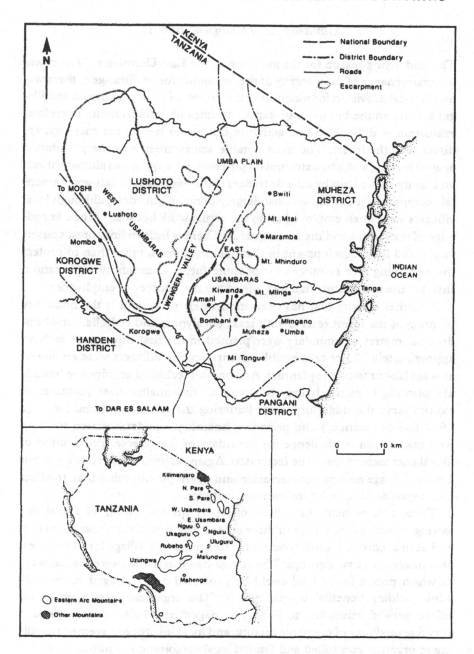

Figure 6.2 The East Usambara Mountains, Tanzania: location map (adapted after Hamilton & Bensted-Smith 1989).

Maintenance of biological diversity

The underlying reason for the initiation of the East Usambaras project was the maintenance of the integrity of the remaining forest. Strangely, there was no practical provision for monitoring the impact of project activities and illegal actions on the biodiversity status and rates of deforestation. Therefore, evaluation is difficult. Some activities but by no means the majority, are directed at the forest. The most notable achievement was the prohibition against logging and a moratorium on pit-sawing. Logging was almost entirely an activity carried out by outside traders and a parastatal (quasi-governmental) company. Pit-sawing was also largely done by outsiders, although local villagers were often employed to do the actual work because of their knowledge of the terrain and the forest guards. There is little to link forest conservation and local development in these achievements, other than to protect the remaining tree resources for possible future community exploitation. Instead, the community had lost one of the few sources of employment.

A further activity that has been moderately successful is the boundary planting of the forest reserve with teak, eucalyptus and cedrella. One hundred kilometres of boundary were planted in the first three years with an approximately 50 per cent establishment success. Villagers were employed as wage labour to do the planting. As a one-off technical activity, the boundary planting is relatively easy to organize. Essentially, these plantings of exotics serve the dual purpose of buffering the remnant natural forest (a "first line of defence") and providing unambiguous demarcation to those who might wish to challenge the boundary of the reserve. Prosecution of illegal encroachers would be facilitated. Again, in itself, this activity was not a direct linkage between conservation and development, other than to effect a surveyed basis for a future management plan.

These achievements have been offset by the flourishing of illegal pit-sawing, as well as the failure of three other forest conservation activities that had been accorded priority: communal tree nurseries, village forest reserves, and village pit-sawing groups. The village-based activities were seen as ways in which public forest land could be protected and managed sustainably while yielding benefits to local people. The organization of such group efforts proved frustrating to project management: the nurseries provided very few seedlings of plantable quality, and local institutions seemed unwilling to organize controlled and limited legal exploitation of public land.

The disappointments in addressing biodiversity issues directly suggest some practical aspects that should be considered in any attempt at hitching conservation objectives to development. First, new regulations (e.g. the moratorium on pit-sawing) are almost impossible to enforce, especially where they withdraw previous rights of access. Alienation of the project and its activities from influential persons in the village is bound to result. Secondly, most conservation objectives cannot be achieved within the standard project planning period of 3–5 years. Buffer zones, village forest reserves and the like will take at least a decade to become accepted; a conservation ethic may take at least a generation to develop. Progress on communal activities, where the private benefit is subjugated to the common good, is similarly bound to be slow. The institutions to organize such developments have to grow organically within local society; they cannot be externally forced.

Resolution of land-use conflicts

By far the greatest number of project activities concern agricultural improvements, infrastructural and other developments, research and training. The primary aims of all such work is to:

- divert attention away from the forest
- protect land that is already deforested and probably already degraded
- increase productivity of land outside the forest reserves in order to increase living standards.

The potential conflict between people and reserve is therefore addressed by a combination of "keep them busy", "keep them happy" and "keep them out [of the forest]". What is the record to date?

Cardamom production within the forest was one of the major threats to natural forest regeneration. This is tackled by promoting cardamom in an agroforestry system on farmers' fields, and other cash crops such as pepper, cloves, coffee and cinnamon to replace lost income. The difficulty of tackling cardamom is that it is mainly grown by estate workers for a subsidiary income and by the landless. The neat association between forest conservation through agricultural improvement is confounded, because those who benefit by the development are not those who suffer from the withdrawal of access to the forest to grow cardamom. Then, if cardamom production were to expand on farmers' fields, producer prices would decrease. Those who grow cardamom illicitly in the forest would then have to increase their area of planting to make up for lost income, thereby having the very opposite impact than was intended!

Soil conservation was another major effort. A simple contour-pegging device, the "Amani" level, was designed, tested and used by farmers; planting materials of sugar cane, Guatemala grass and pineapple were provided for permanent vegetated contour banks. There was considerable initial enthusiasm over receiving and operating the levels, and obtaining free plants. However, by 1990, activity was waning: 15 per cent of fields were pegged, only half of these had been contour planted, and, most revealing of all, almost no repairs had been made to breached contour lines. Clearly, farmers were less enthusiastic about the efficacy of the technique than they were to receive attention by the project staff.

In terms of resolving potential conflicts between the forest and the people, the project was at best benign and at worst malign to the goals of conservation.

Lessons from the Usambaras in integrating conservation and development

The East Usambaras Agricultural Development and Environmental Conservation Project is at heart a conservation-driven activity: the aim is to secure the forest. From a local person's perspective it could easily be perceived as "conservation-by-stealth", where protection of the forest is effectively bought at a developmental price. The vast majority of the project's activities are outside the forest and inside the communities that might otherwise threaten the forest. Here is the primary link between conservation and development: a conservation objective by developmental means. Is this ethical, practical, sustainable?

The ethics are perhaps questionable. First, the ethical arguments of conservation discussed earlier in this chapter of the intrinsic worth of species to exist and of the duty of humankind of stewardship of the natural world are difficult to sustain alongside an ethic of human dignity and the right of human beings to determine their own future. Secondly, there are the ethics of creating a hidden agenda, the objective of conservation, behind a cloak of economic development. Is it acceptable to put conditionalities on development, a sort of environmental blackmail, in order to force people to respect a resource? On the one hand, some would agree that the ends justify the means and that people need coaxing, even to protect their own future. On the other hand, many would balk at the implication that society should underwrite the protection of biodiversity through funding development services. These ethical dilemmas are part of the difficulty of balancing the different perspectives of the value of biodiversity (see Table 6.2). The consumptive use values that local people give to the forest are vastly different to the production use values of the local market, which in turn are different from the non-consumptive use values of local environmentalists and the option values and existence values of the conservation community. The project begs the question as to whose ethics are paramount, and can the moral arguments for biodiversity ethically be bought with money?

Practicality is also in question. Aid donors, governments and villagers themselves insist on the delivery of quick, tangible benefits. Promises of long-term benefits are less real, uncertain to quantify, and extremely risky to achieve, when other variables are taken into account, such as wars, coups, unpredictable bureaucracy and the vagaries of international trade. To a local villager, conservation may be perceived to be the price as paid for otherwise unavailable development funds, economic development being the reality now and conservation-based improvement an insecure wishful thought.

The Usambaras case also highlights lessons about sustainability. The ecological presentation of sustainability, as described in *Caring for the Earth*, is unambiguous: attention to soil erosion, promotion of agroforestry, the integration of diverse land-use activities – all have a sound rationale. However, it is the way that the ecological sustainability can be delivered in the future that is potentially unsustainable. Many projects devote huge resources, experienced and dedicated staff, and lots of attention to local people. Under these circumstances any reasonably sound technique would probably be implemented by villagers, proud of the attention, kudos and additional opportunity such a project brings. But to reproduce that intensity of project activity beyond pilot areas is impossible. Even in the project areas, once the aid has pulled out, doubts must remain whether project achievements will still be respected.

Lest these views be thought unduly pessimistic, there are some positive lessons to be learnt. Conventionally trained conservationists are now entering the real world of humanity. The understanding of why people, usually poor and marginalized, react in the way they do to undermine resources, has grown apace; the political economy and political ecology of natural resource use is a legitimate area of study. A related positive aspect is the definite increase in participation by local communities. Village co-ordinators, elected by villagers themselves, have fostered a climate of respect and co-operation, and evaluation missions of the Usambara project have commended staff for the degree of local involvement in all activities. These are small but significant steps upon which to build.

Conclusion

This chapter opened with the perceptions of a conflict between nature and the use of natural resources, and argued that biodiversity is worth protecting for its many values to society. So valuable is biodiversity that tracts of productive land are set aside for protection. However, there is evidence from many sources that, despite apparently logical reasons, the reaction of conservationists has sharpened the conflict by separating conservation into protected areas and expelling people to intervening localities. Thus, it has been revealing to see the attempts to reverse that process, primarily from conservation organizations.

Reversal has been attempted mainly by linking conservation goals and actions with development, by arguing that one is contingent upon the other, by showing the "true" value of conservation and what it can contribute to development, and by developing a more flexible approach to protected areas. This chapter has looked at two such initiatives: a multiple-use area and a managed forest. The means of linkage are usually through externally funded projects, institutions and government initiatives. The overall conclusions may be summarized as:

- **Local participation** is essential in conservation and development. Without it, the very people who are expected to respect nature are excluded from planning and implementation – an impossible situation to sustain;

- Integrated conservation and development projects are extremely **complex and difficult to manage**; they involve a delicate balance between conservation goals and development priorities. If community participation is also included, then conflicts between conservation and development are bound to arise. In the Ngorongoro case, participation by the Maasai was minimal, and conservation interests were overwhelmingly dominant, to the detriment of local people. In the Usambaras case, most activities were development-orientated but still required much longer than a typical project cycle.

- Conservation and development need **explicit linkage**. It requires a far better understanding of socio-economic issues than conservationists normally hold; and better appreciation of unconventional and flexible responses required for conservation by development professionals. Development activities are not always compatible with conservation objectives, even when designed by the same team. Only mutually supportive activities should be subsidised from external sources;

- A **change in attitudes** towards more sustainable living on the part of all actors in conservation and development is required. Training, education and research through monitoring of activities are all part of the learning process. This must be a long-term aim towards which short-term inputs through projects should seek to make a contribution.

A few years ago we stated that there was no really viable alternative but to link conservation with development, if biological diversity is to be protected and if local people are to retain a control over their resources through sustainable development (Stocking & Perkin 1992). We have no reason to change our view. However, in order to coexist with nature in a developing world, many more attempts will be needed. These attempts should recognize the diversity of environments and of peoples. There are no universal solutions, no panacea. Solutions will have to be built upon the resolution of conflict rather than on defensive positions for either conservation or development. Multiple-use areas and local control of reserves would seem to be two useful models, but ones that need much more planning, management and empathy than they have received hitherto.

References

Armstrong, S. J. & R. G. Botzler (eds) 1993. *Environmental ethics: divergence and convergence.* New York: McGraw-Hill.

Bell, R. H. V. 1987. Conservation with a human face: conflict and reconciliation in African land use planning. In *Conservation In Africa*, D. Anderson & R. Grove (eds), 79–101. Cambridge: Cambridge University Press.

Biot, Y., R. Lambert, S. Perkin 1992. *What's the problem? An essay on land degradation, science and development in sub-Saharan Africa.* Development Studies Discussion Paper 222, School of Development Studies, University of East Anglia.

Bower, J. R. F. & P. Gogan-Porter 1981. *Prehistoric cultures of the Serengeti National Park: initial archaeological studies of an undisturbed African ecosystem.* Papers in Anthropology 3, Department of Sociology and Anthropology, Iowa State University, Ames.

Broad, R. 1994 The poor and the environment: friends or foes? *World Development* 22, 811–22

Brown, K. 1994. Approaches to valuing medicinal plants: The economics of culture or the culture of economics? *Biodiversity and Conservation* 3(8), 734–50.

Brown, K. & D. Moran 1994. Valuing biodiversity: The scope and limitations of economic analysis. In *Biodiplomacy: genetic resources and international relations*, V. Sanchez & C. Juma (eds), 213–32. Nairobi: ACTS Press.

Brown, K. & D. W. Pearce (eds) 1994. *The causes of tropical deforestation.* London: UCL Press.

Government of Tanzania 1975. *Game Parks Laws (Miscellaneous Amendments) Act.* Dar-es-Salaam: Government Printer.

Guha, R. 1989. *The unquiet woods: ecological change and peasant resistance in the Himalaya.* Delhi: Oxford University Press.

Hamilton, A. C. & R. Bensted-Smith (eds) 1989. *Forest conservation in the East Usambara Mountains.* Gland, Switzerland: International Union for the Conservation of Nature and Natural Resources.

IUCN 1985. *Threatened protected areas of the world.* Gland, Switzerland: International Union for the Conservation of Nature and Natural Resources.

IUCN/UNEP/WWF 1980. *World conservation strategy: living resource conservation for sus-*

tainable development. Gland, Switzerland: International Union for the Conservation of Nature and Natural Resources.

— 1991. *Caring for the Earth: a strategy for sustainable living.* Gland, Switzerland: International Union for the Conservation of Nature and Natural Resources.

Juma, C. 1989. *The gene hunters: biotechnology and the scramble for seeds.* London: Zed Press.

Ledec, G. & R. Goodland 1988. *Wildlands: their protection and management in economic development.* Washington DC: World Bank.

Lovejoy, T. 1980. Projections of species extinctions. In *The Global 2000 report to the President,* G. Barney (ed.), 328–31. Washington DC: Council on Environmental Quality.

MacArthur, R. H. & E. O. Wilson 1967. *The theory of island biogeography.* Princeton, New Jersey: Princeton University Press.

Machlis, G. E. 1992. The contribution of sociology to biodiversity research and management. *Biological Conservation* **62**, 161–70.

Maro, W. E. 1990. *Education and health services within the Ngorongoro Conservation Area.* Technical Report 13, Ngorongoro Conservation and Development Project, IUCN Regional Office for Eastern Africa, Nairobi.

McCabe, J. T., S. Perkin, C. Schofield 1992. Can conservation and development be coupled among pastoral people? An examination of the Maasai of the Ngorongoro Conservation Area, Tanzania. *Human Organization* **51**, 353–66.

McNeely, J. A. 1988. *Economics and biological diversity: developing and using economic incentives to conserve biological resources.* Gland, Switzerland: International Union for the Conservation of Nature and Natural Resources.

McNeely, J. A., F. R. Miller, W. V. Reid., R. A. Mittermeier, T. B. Werner 1990. *Conserving the world's biological diversity.* Gland, Switzerland: International Union for the Conservation of Nature and Natural Resources.

Mturi, A. A. 1990. *Archaeological and palaeontological issues in the Ngorongoro Conservation Area.* Technical Report 12, Ngorongoro Conservation and Development Project, IUCN Regional Office for Eastern Africa, Nairobi.

Myers, N. 1979. *The sinking ark.* New York: Pergamon.

— 1984. *The primary source.* New York: Norton.

Naess, A. 1989. *Ecology, community, and lifestyle.* Cambridge: Cambridge University Press.

Nestel, P. 1986. A society in transition: developmental and seasonal influences on the nutrition of Maasai women and children. *Food and Nutrition Bulletin* **8**, 2–18.

Pearce, D. W. 1993. *Economic values and the natural world.* London: Earthscan.

— 1994. The great environmental values debate. *Environment and Planning A,* **26**, 1329–38

Pearsall, S. 1984. *In absentia* benefits of natural preserves: a review. *Environmental Conservation* **11**, 3–10.

Peluso, N. 1993 *Rich forests, poor people: resource control and resistance in Java.* Berkeley: University of California Press.

Pimentel, D. (ed.) 1993. *World soil erosion and conservation.* Cambridge: Cambridge University Press.

Raven, P. H. 1988. Our diminishing tropical forests. In *Biodiversity,* E. O. Wilson & F. Peters (eds), 119–22. Washington: National Academy Press.

Reid, W. V. 1992. How many species will there be? In *Tropical deforestation and species extinction*, T. C. Whitmore & J. A. Sayer (eds), 55–74. London: Chapman & Hall.

Repetto, R. & M. Gillis 1988. *Public policies and the misuse of forest resources.* Cambridge: Cambridge University Press.

Rodgers, W. A. & K. M. Homewood 1982. Species richness and endemism in the Usambara mountain forests. *Biological Journal of the Linnean Society* 18, 197–242.

Rolston III, H. 1988. *Environmental ethics: values in and duties to the natural world.* Philadelphia: Temple University Press.

Spellerberg, I. F. & S. R. Hardes 1992. *Biological conservation.* Cambridge: Cambridge University Press.

Stocking, M. A. (in press). Soil erosion. In *The physical geography of Africa*, A. S. Goudie, W. M. Adams, A. Orme (eds). Oxford: Oxford University Press.

Stocking, M. A. & S. Perkin 1992. Conservation-with-development: an application of the concept in the Usambara Mountains. *Institute of British Geographers, Transactions* 17, 337–49.

Tiffen, M., M. Mortimore, F. Gichuki 1994. *More people, less erosion: environmental recovery in Kenya.* Chichester: John Wiley.

Timberlake, L. 1985. *Africa in crisis: the causes, the cures of environmental bankruptcy.* London: Earthscan.

Wells, M. 1992. Biodiversity conservation, affluence and poverty: Mis-matched costs and benefits and efforts to remedy them. *Ambio* 21, 237–43.

— K. Brandon, L. Hannah 1992. *People and parks: linking protected area management with local communities.* Washington DC: World Bank.

Wilson, E. O. 1988. The current state of biological diversity. In *Biodiversity*, E. O. Wilson & F. M. Peters (eds), 3–20. Washington: National Academy Press.

Wilson, E. O. & F. M. Peters (eds) 1988. *Biodiversity.* Washington: National Academy Press.

World Bank 1992. *Development and the environment*, World Development Report. New York: Oxford University Press.

CHAPTER SEVEN

The next 1000 million people: do we have a choice?

Ian Thomas

Editors' introduction

Development is all about people, and a chapter on population is a must in any book of this type, which examines current issues and debates in the complex field of development. Population and development have been linked so often that one is now often regarded as the mirror image of the other. It is often assumed that high population growth rates means underdevelopment and environmental degradation, implying a clear cause–effect relationship in the minds of many. Indeed, in an opening speech at the Rio Earth Summit, Maurice Strong, the Secretary-General of the Summit, highlighted the "explosive increase in population" as a major environmental hazard. However, although population is mentioned in many of the Rio declarations, increases in the world's population was not tackled directly as a central element. Early drafts of the Agenda 21 document did propose linking poverty, overconsumption of resources in rich countries, and rapid population increases in the Third World, but some Northern governments did not like the inclusion of overconsumption. As a response, several developing countries refused to allow discussion of population control at Rio. The preference instead was to leave such discussions until the Cairo Population Summit of 1994.

Population did indirectly enter various Rio declarations, such as those dealing with agriculture, land reform and access to water, and some of these facets are dealt with in other chapters of this book. Relevant questions include:

- How can the demands of more people be met in tune with the need to conserve biodiversity (Ch. 6)?
- Will farming be able to sustainably produce more in order to feed the ever increasing number of mouths (Ch. 2)?
- Will a technical fix such as biotechnology come to the rescue (Ch. 5)?

A further assumption is that the human population is set to expand rapidly in the coming years. This fuels the fears of many who see the threat to the environment rapidly assuming tangible form. As politicians and planners appear to be faced with the inevitable, this chapter examines what is happening to population numbers, what has produced rapid growth and asks whether it is as inexorable as it appears.

Introduction: do we have a choice?

The UN estimated the world population in mid-1990 as 5300 million (UN 1991). They estimate future population size with different sets of assumptions about fertility and mortality change, and this gives different rates of growth. In 1991 the medium variant projection resulted in a mid-year 2000 population of 6300 million. That is an extra 1000 million within a decade. Other writers in this volume are reviewing the future with this and subsequent massive population growth necessarily in mind; part of their task has been to consider the implications of such growth. It is the purpose of this chapter to examine why this amount of growth is occurring, where it is likely to be greatest, what prospects there might be for different outcomes, and what factors influence these prospects.

Some of the demographic changes during recent decades have been dramatic. Fertility reduction in China is a frequently cited case (Peng Xizhe 1991), as is that of Kerala in south India (Alam & Leete 1993). The mortality decline of Sri Lanka (Langford & Storey 1993) is equally well known, as is that of Nicaragua (Garfield & Williams 1989). Table 7.1 shows some of these demographic changes. Population forecasts are notoriously unreliable, because analysts, it is argued, fail to foresee such new demographic directions. However, perhaps circumstances may change during the next few years in such a manner that it will take far more than a decade for the next 1000 million people to be added to the world's population? It seems highly unlikely within the demographically short timespan of ten years. What may be possible is that the prospects for more moderate increases in numbers during the twenty-first century are improved during this decade. Global warming and its consequences, disruption of the ozone layer, the spread of HIV/AIDS or some other deadly virus, or wars of an even greater scale, may increase mortality and thereby diminish population growth, but that is hardly an improvement. The preferred route to lesser growth, if there is to be a reduction, must be decline in fertility.

The choices available may be examined from both empirical and theoretical standpoints. Countries of the world currently exhibit a wide range of population conditions, even when compared with a few simple demo-

Table 7.1 Rapid demographic change: selected examples.

	China	Sri Lanka	Nicaragua	Kenya	Less developed[a]
Population (millions)					
1950	554.8	7.7	1.1	6.3	1684
1970	830.7	12.5	2.1	11.5	2649
1990	1139.1	17.2	3.9	24.0	4086
Exp. of life at birth, both sexes (years)					
1950[b]	40.8	56.6	42.3	40.9	42.2
1970	63.2	65.0	54.7	51.0	55.2
1990	70.9	71.6	66.3	61.0	63.3
Total fertility rate (births per woman)					
1950[b]	6.24	5.74	7.33	7.51	6.19
1970	4.76	4.00	6.71	8.12	5.41
1990	2.25	2.47	5.01	6.80	3.71

Source: UN 1991, pp. 230–1, 354–5, 450–1, 506–7, 556–7
Notes: (a) All regions of Africa, all regions of Asia excluding Japan, Melanesia, Micronesia and Polynesia.
(b) Estimates for the five-year periods 1950/55, etc.

graphic indicators such as their population growth rates, their expectation of life at birth, and their family size (total fertility rate, TFR, is the estimated number of births per woman). It is interesting to examine the influence nations have on the national growth paths and composition of their own population. Have they made explicit choices or is their demographic condition the outcome of wholly external forces such as the international programmes of disease control? In what sense may nations be said to choose the aggregate effect of individual sexual behaviour, and the chances of death? Demographic models have been developed (Coale & Demeny 1983) that permit examination of the long-term outcome of sustaining particular demographic choices: there are stark contrasts in the demographic parameters and resulting population structures (Fig. 7.1).

The age distributions in Figure 7.1 represent stable and stationary populations. When populations experience fixed age distributions of deaths and fertility over a long period, they establish an unchanging age distribution: the proportions at each age become constant. This is a stable population. If the mortality and fertility rates are such that the resulting rate (called the intrinsic rate) of population growth is zero, then the number at each age remains the same, as well as the proportion. This is a stable and stationary population. In one sense, the pyramids in the figure represent extremes. Case A has very low expectation of life at birth and relatively high fertility. Case B has high expectation of life at birth, but to remain stationary has low total fertility. This highlights one demographic choice that might be made: mortality may be reduced, but to keep a check on population growth, fertil-

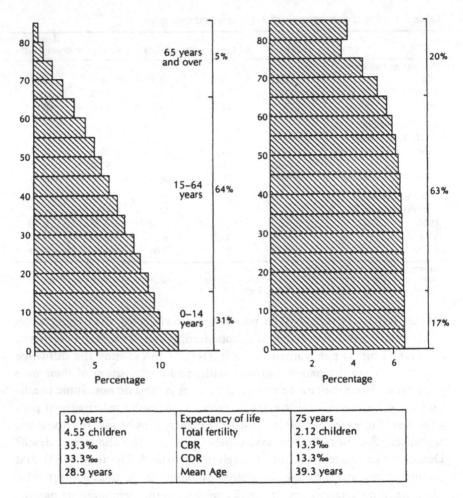

30 years	Expectancy of life	75 years
4.55 children	Total fertility	2.12 children
33.3‰	CBR	13.3‰
33.3‰	CDR	13.3‰
28.9 years	Mean Age	39.3 years

Figure 7.1 Transition: a conscious choice? Two stationary populations (females).

ity must also be reduced. With low mortality and fertility, and population stable and stationary, the mean age is high; this is not so much a choice as a consequence. If a young population is sought under stable and stationary conditions, it results from the high mortality/high fertility option.

In an important sense, the demographic values and population compositions in Figure 7.1 are not extremes. Fertility rates are found in many countries at much higher and lower values than those used in the stationary cases. Also the mean ages of actual national populations vary more: pyramids, in reality, may be more broadly based and "young", or more top heavy and "old". However, no national population is stable or stationary. No country has experienced unchanging mortality and fertility schedules

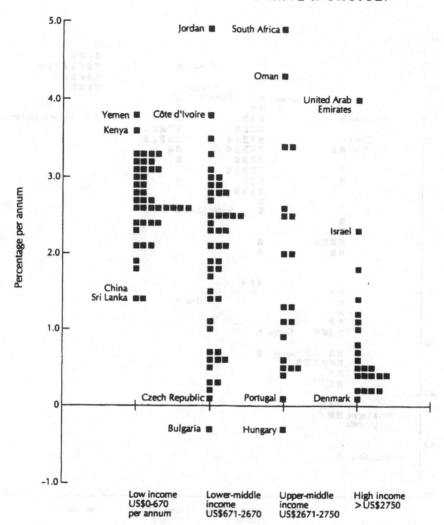

Figure 7.2 National population growth (% per annum), 1980–92 (by World Bank income groups). *Source: World development report*, World Bank (1994: table 25).

throughout the past 100 years or so. Consequently, there is a range of national population growth rates from less than zero to 4 per cent per annum or more (Fig. 7.2). Population compositions and size also reflect the demographic experience of decades of change. Demographic rates, similarly, represent the results of this dynamism (Figs 7.3, 7.4).

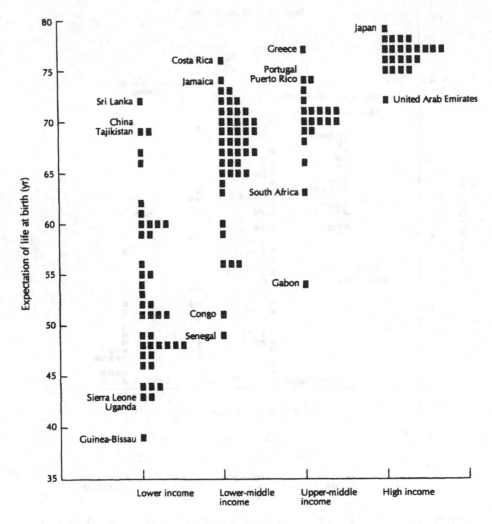

Figure 7.3 Expectation of life at birth, 1992 (by World Bank income groups). *Source: World development report*, World Bank (1994: table 7).

Trend in mortality

All nations seek to increase life expectation at birth for their citizens, whether the issue receives explicit treatment in their national development plans or not. In this they have been notably successful. The United Nations Population Division has recently produced estimates of life expectancy for developed and developing countries for the period 1950–55 to 1980–85

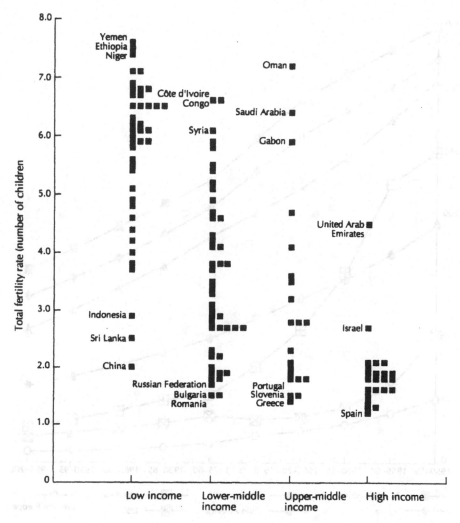

Figure 7.4 National fertility, 1992 (by income groups). *Source: World development report,* World Bank (1994: table 26).

(UN 1989a: 22). In developed countries it has increased for females from 68.7 years to 76.9 years, and for developing countries from 41.9 years to 58.3 years. Similar changes are recorded for males, although at somewhat lower levels: 63.0 to 69.5, and 40.3 to 56.3 years. This is a remarkable performance. Few other development indicators could be found to show such sustained progress over that time, and with a greater rate of improvement in developing than in developed countries.

The record with mortality of children under five is even more astonishing (Fig.7.5).

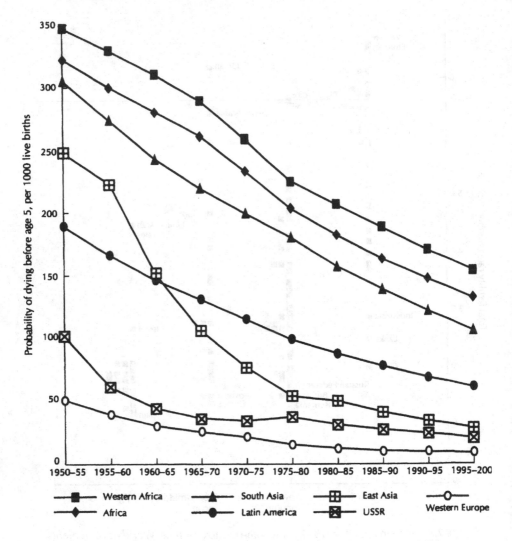

Figure 7.5 Under-five mortality, 1950/55 to 1995/2000 (*Source:* UN 1988).

There are still major disparities between the geographical regions, but even the most disadvantaged have achieved major advances in reducing the probability of deaths in young childhood. The case of East Asia shows what can occur. All nations, one might suppose, would choose to reduce childhood mortality, but some have been significantly more successful at doing so than others. One may assert with a reasonable degree of confidence that this is the exercise of choice.

Means of achieving mortality decline

Shortly after Dr Hiroshi Nakajima became Director General of WHO in 1989, he described as "the silent genocide", child deaths from six major infectious diseases: polio, tetanus, measles, diphtheria, whooping cough, and tuberculosis (*The Observer*, 1 October 1989). The basis of this emotive statement was that, for an average cost of only £6 a head, children could be vaccinated against these diseases, and most of the 11 million or so child deaths that occur annually from these causes were preventable at relatively low expense. UNICEF has made a mission of extending immunization programmes throughout the world and reducing the level of child mortality (UNICEF, annually). Its campaign combined with variable inputs from national governments has had considerable success, but has also highlighted problems about health-care delivery that are not always easy to overcome. Medical technology has produced effective methods of limiting the incidence of infection and it has also found ways of providing vaccines at relatively low cost. Persuading governments to accept and fully promote immunization campaigns, and mounting delivery systems that make vaccination acceptable and accessible to communities, has proved very difficult in some countries. Variations in the extent to which these problems are surmounted accounts largely for the different degree of success in combating child mortality from infectious disease among the nations.

Attempts have been made to isolate the characteristics of nations that are associated with successful mortality reduction, and those that are lagging. Caldwell (1986) identified what he categorized as "superior" and "poor" health achievers by the simple device of comparing world ranking by per capita income and by expectation of life at birth and infant mortality rate. Countries that had better mortality status than their per capita wealth would suggest included Sri Lanka, China, Myanmar (Burma), Jamaica, India, Zaire and Tanzania. Those in which mortality appeared worse than one might expect from the level of per capita income included Oman, Saudi Arabia, Iran, Libya, Algeria, Iraq, Yemen Arab Republic, Morocco, Ivory Coast, Senegal and Sierra Leone. He then went on to examine in detail the evolution of health policy and programmes in Sri Lanka, Costa Rica, and Kerala in south India, in order to explain the success they had achieved in improving the health of their populations. He found it necessary to extend his assessment to features of the society not normally regarded as part of the health system: political history, cultural practices, educational provision and administrative efficiency all appeared as important components of increasing the length of life of citizens. Poor health achievement, in the terms he used, is to be explained by inadequate health systems, but also by urban and elite bias in the availability of a wide range of services, notably education, and by lack of female autonomy in most spheres of life. Superior health achievement was associated with a moderate to high degree of female freedom and opportunity to act independently, a willingness to channel public resources into both health and education that was then made accessible to males and females, and an efficient health service. The latter was characterized by attention to preventive as well as curative activities, and an effort to make effective use of resources at the local level by strengthening a primary health care focus on maternal and child health care that included universal immunization, and the maintenance of adequate nutrition for all, even for the most economically disadvantaged. These features were most likely to be present when political and cultural circumstances stimulated a popular demand for basic health care, and this then translated into sustained political commitment to the provision of social services, irrespective of changes of government.

Clearly, choices had been exercised by leaders, governments and peoples that resulted in improved health and longer life. However, successful outcomes emerged from sustained effort across a wide range of political and social service activities. Economic strength did not appear to be a major determinant; although adequate resource provision was essential, resources alone are not sufficient. No political system appeared to have a monopoly of success; superior and poor achievers exist under both capitalist and socialist systems. Since 1990 in its annual publication, the *Human development report*, the UNDP has been highlighting the choices open even to poorer nations, and promoting programmes similar to those identified by Caldwell as a means of improving the welfare of poorer people in all societies. Switching expenditure from military purposes to selective social welfare (primary education, and primary health care for instance) is one such shift of emphasis. If successful, these programmes are likely to boost population growth by their reduction of mortality, at least in the short to medium term of 5–15 years.

Trend in fertility

National population growth is the outcome of mortality, fertility and net migration (the balance of emigration and immigration). With a few exceptions, the key determinants of national population growth in recent decades have been mortality and fertility. They are the only determinants of global population change. Although, as we have seen, rates of decline have varied among nations, mortality has fallen substantially during the past 40 years and is likely to continue to do so for some time to come. During the 1990s, changes in mortality will almost certainly continue to contribute steadily to greater population growth.

Whether it is rapid, very rapid or just moderately high, the rate of population growth during the next decade or two depends to a large extent on what happens to fertility. The past 40 years of change in fertility among the nations is more complex than that of mortality (Fig. 7.6).

Some major world regions, notably the Far East again, have experienced steep and sustained decline in average family sizes. Others, notably Africa but also parts of South Asia, appear to have had fertility increases or have experienced plateaux of high fertility.

Figure 7.6 Total fertility rate, by income groups and selected regions, 1970–2000. *Source: World development report,* World Bank (1994: table 26).

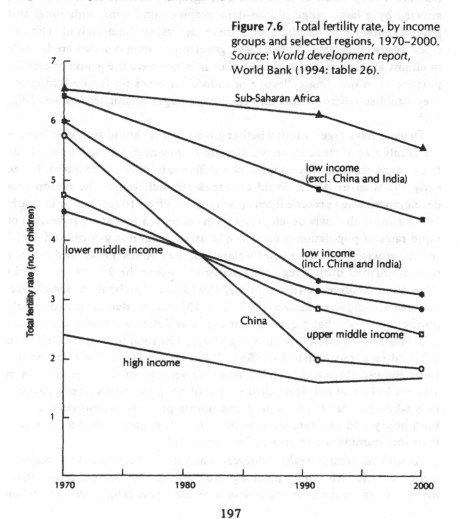

Explanations for fertility level and trends

Changes in total fertility rates have been attributed to four sets of proximate demographic determinants: the amount of celibacy, the number of induced abortions, the prevalence of contraception, and the extent of sterility (Bongaarts & Potter 1983). Celibacy includes delay before first marriage as well as periods of divorce and widowhood. Sterility includes primary and secondary forms, the first being the 3 per cent or so of individuals who are incapable of reproduction, and the latter covering a range of conditions that limit fertility once a birth has occurred (some temporary, such as the short infecund period immediately after a birth, and others permanent; Larsen & Menken 1991). Each of these four demographic determinants is, in turn, affected by a large range of non-demographic conditions. Individual and governmental decisions affect all of these factors, so the notion of "choice" is less easily applicable. Nonetheless, government interventions are feasible in almost all aspects and, although they may not have the prime or explicit purpose of influencing collective or individual views on family formation, they establish environments that encourage larger or smaller families (Fig. 7.7).

Debates have raged about whether governments should seek to influence the family size of citizens, on what grounds they might do so, the methods to be employed, and why couples have different numbers of children. In the early 1970s many Third World countries still adhered to the notion that development must precede family planning (Wolfson 1978), but by the early 1980s almost all newly developing nations accepted that some reduction of rapid rates of population growth would assist economic growth and social service provision (Kilimanjaro Declaration, see UNECA 1984), and this has been confirmed during regional discussions before the 1994 Cairo World Population Conference (for example, 1992 Dakar Declaration, see *Population and Development Review* 1993: 209–15). China demonstrated clearly during the 1970s that a government that was determined to reduce births could do so on a massive scale (Peng 1991). The total fertility rate fell from 6.2 children per woman in 1950/55 to 2.4 in 1980/85 (UN 1989a). However, India showed during the period when an emergency was declared from 1975 to 1977 that political will alone could not guarantee success, particularly when policies led to widespread undue pressure on individuals, and both policy and pressure were associated with central government rather than with community interests (Gwatkin 1979).

As with successful health achievers, analysts have identified the requirements for effective family planning programmes by examining countries where contraceptive prevalence has increased and family size has fallen

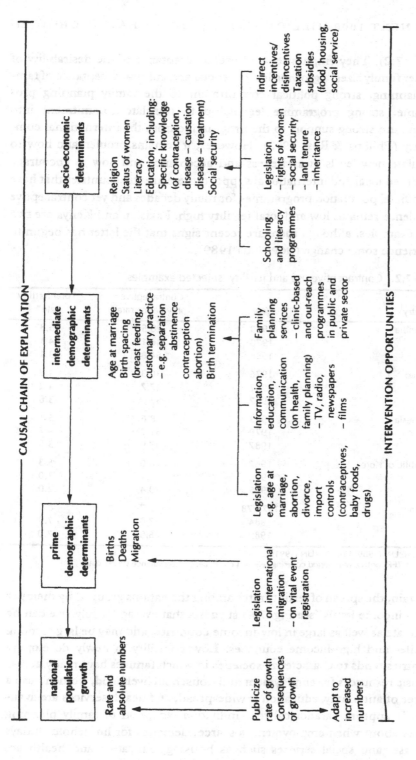

Figure 7.7 The range of possible interventions to influence population growth (Thomas et al. 1987).

(Table 7.2). They emphasize widespread acceptance of the desirability of smaller family sizes, a growing level of social and cultural acceptance of family planning, strong political commitment to the family planning programme, strong programme leadership to translate commitment into action, and strong support to the programme from the international community (Phillips & Ross 1992). However, it remains problematic how to stimulate new levels of awareness and acceptance and how to encourage continued local and international support: there are still countries that have had official population programmes for many decades and yet contraceptive prevalence remains low and total fertility high. Pakistan and Kenya are two such examples, although there are recent signs that the latter has begun to experience some change (NCPD/IRD 1989).

Table 7.2 Contraceptive use and fertility: selected examples.

Country	Year	Contraceptive prevalence rate	Total fertility rate*
Columbia	1969	20.5	6.3
	1978	46.1	4.1
	1986	64.8	3.1
Mexico	1976	30.3	4.9
	1982	47.7	4.2
	1987	53.0	3.6
Indonesia	1973	8.6	5.1
	1979	31.2	4.7
	1987	47.9	3.5
Republic of Korea	1967	20.0	5.3
	1976	44.2	3.0
	1985	70.4	2.0
Kenya	1977/78	6.7	8.1
	1984	17.0	7.9
	1989	26.9	7.0

Sources: UN (1989: 116–26), UN (1991).
Note: * TFR for five-year period covering relevant contraceptive prevalence survey date.

Again, the spread of fertility rates among the nations grouped by their per capita income levels (see Fig. 7.4) suggests that average family size can be moderate as well as large in low-income countries, and may be large among middle- and high-income countries. Lower fertility in newly developing countries tends to characterize societies in which families have ready access to basic resources for employment and household livelihood, women have a degree of autonomy, education is widespread, and health services are available. It appears paradoxical that individual adoption of family planning comes about when employment is scarce, demands for household outlays increase, and social services such as housing, education and health are under some pressure. Knowledge and opportunity must exist, so that real needs may be translated into effective action; couples need to know that it is possible to control fertility, have the motivation to do so, and methods are accessible to them. Governments can facilitate this by having exercised a prior choice and supported information, education and social welfare provision programmes so that when couples wish to exercise their choices the means are at hand (Lapham & Simmons 1987, Phillips & Ross 1992).

Options for the future

Medium variant projections for world population during the 1990s suggest that there will be approximately 1500 million births and 500 million deaths giving a net increase of 1000 million between 1990 and the year 2000. The world's population is increasing at the rate of approximately 1000 million per decade (Table 7.3).

Table 7.3 Births, deaths and population size (millions) in the 1990s (UN medium variant).

	More developed 1990 – 2000		Less developed 1990 – 2000		World total 1990 – 2000	
Births	174		1308		1482	
Deaths	116		397		513	
Balance	58		911		969	
Population	1207	1265	4085	4996	5292	6261

Based on UN (1991: 100, 226–31). UN assumes net migration is zero.

Deaths could be greater if AIDS, war, famine or natural catastrophe occurs on a greater scale than was anticipated during the last few years of the 1980s when the UN projections were prepared. The evidence on AIDS is conflicting, and knowledge is fragmentary, but the impact on death rates within the next few years is not likely to be great by comparison with other major causes of death such as malaria, cancer, heart disease or accidents (Garnett & Anderson 1993, World Bank 1993). The other positive Malthusian checks to population increase remain less predictable, and are generally ignored for the purposes of calculating population projections. In the longer run, global warming or destruction of the ozone layer might produce major increases in mortality (see Ch. 3). In its annual *World development report*, the World Bank has for the past few years been publishing estimated years at which national populations will attain a net reproduction rate (NRR) of 1 (that balance of fertility and mortality schedules, referred to at the outset of this chapter and which – if sustained – would eventually produce a stable and stationary population), and a hypothetical constant population size. NRR of 1 is estimated to be reached from the present through to the year 2055, and this would give a stable world population approximately 70 years later of 11/12 000 million, with major consequences for the redistribution of world population (Fig. 7.8). Within this timespan major changes in mortality could occur, but there is no reason to assume that they will do so within

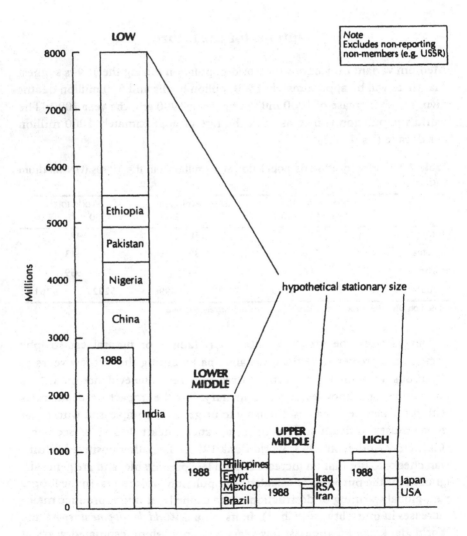

Figure 7.8 Population by World Bank income groups (1988) and hypothetical stationary populations. *Source: World development report*, World Bank (1994: table 26).

a decade or that nations will change their basic predilection for a longer life.

The future direction of fertility change seems similarly as predictable. Global fertility is set to decline. However, the pace of decline, and the path of change for individual nations is much less certain. Many industrialized countries now have average fertility at unprecedentedly low levels: well under the replacement level of 2.1 TFR (see Fig. 7.4). This may rise. Even if it does, the impact on global population numbers is relatively small, because industrial country births make up a comparatively small proportion of all births. Analysts were unwilling to accept evidence of fertility increase in developing countries during the 1960s and early 1970s, but examination of the components of change in average total fertility, combined with better demographic data collection and analysis, have shown how this may come about (Blacker 1993). These studies raise important issues about the extent to which couples have the power of choice.

Individuals in societies acquire a set of norms and expectations. The impact of some of these is recognized explicitly, such as having a spouse and children, whereas others operate less obviously, such as the effect on ultimate family size of age at marriage, infant and young child deaths, separation of spouses, conjugal abstinence customs following the birth of a child (post-partum abstinence) and breastfeeding. Circumstances affecting the lives of individuals change in ways not always of their own choosing, or, if from choice, with some effects not anticipated. This has happened widely with mortality- and fertility-affecting conditions during the past three to four decades. Medical and environmental conditions (including water supply, drainage, sewerage, waste disposal and pest control) have changed, so that pregnancy wastage has declined, and infant and child mortality has been reduced. The chances have been improved of a conception resulting in a live birth, and a live-born child growing up. Meanwhile, reductions in breastfeeding, erosion of traditional intercourse taboos, and elimination of residential arrangements that formerly separated spouses, have often shortened birth intervals. Conceptions occur more frequently, as well as having a greater chance of producing a grown child. Economic conditions may or may not put a value on children, and the society may or may not regard as a blessing the addition of extra numbers, but for the parents – and particularly for the mother – the "choice" of a larger number of live births and of a larger living family may be more apparent than real.

In model terms, Easterlin (1975) has identified as crucial that point when an excess demand for children becomes an excess supply, and the causes are changes in both the chances of birth and the chances of survival. The fact, as far as it can be established as a fact, that 30 to 55 million abortions occur every year, half of them illegally (Johns Hopkins University 1980, IPPF 1993) suggests that a substantial proportion of all conceptions are unwanted, and surveys suggest that in many high-fertility countries a large percentage of married women of reproductive age want no more births (Robey et al. 1993). The steady increase in contraceptive use is another indication of the desire to increase control over conception and family size (Table 7.3).

The balance of increased probabilities of birth and survival, and increased desire and willingness to control births by traditional or modern methods, means that the future path of fertility is uncertain, even over the next ten years. The effect of varied assumptions on population outcomes can be considerable. Whereas the UN medium-variant assumption for the world's population in 2000 is 6260 million, the constant fertility assumption is 6460 million, the high variant 6420 million, and the low variant 6090 million (UN 1991). The constant-fertility projection assumed the total fertility rate of 1990 remained unchanged until 2000; the high, medium and low variants assumed respectively larger decreases in fertility during the ten years (to TFRs of 3.3, 3.0 and 2.5). When projected for another decade, the varying assumptions themselves produce population estimates with a difference of 1000 million (at 2010: 7900, 7600, 7200 and 6800 million).

Conclusion: demographic outcomes and development

The world's population is continuing to increase rapidly. Our review of the components of that growth, and the proximate and contextual factors affecting it, and likely to affect it over the next decade or two, suggests that development efforts will need to provide for an additional 1000 million people in the next ten years. This includes an assumption that nations and their inhabitants will lower their fertility more than their mortality, thereby reducing their rate of population growth. Increasingly, the means to produce these changes are being made the object of national policy. Citizens are being informed about how to contribute to collective outcomes, and are in turn letting their preferences be known to governments. Individual behaviour is, however, often slow to change, channels of communication are imperfect, and governments differ in the extent of their commitment to improving the welfare of those they govern.

The resulting birth and death rates, combined with net migration effects, will produce varying population growth rates among nations and regions during the 1990s and beyond. This creates differential population growth and shifts in the regional distribution of population, as we have seen. It also contributes to national and regional variations in population composition.

This is not the place for a full analysis of these phenomena, nor are the effects major ones over just ten years, but some of the key characteristics may be noted.

Differential growth rates mean there is an increasing redistribution of the global population from Europe to Africa and Asia, and within Asia from East to South. Rural populations are still building up in absolute number in many Third World areas, but the proportion of national populations in urban areas grows steadily. The contribution of natural increase to the rate of urban population growth may diminish if urban populations respond more quickly to opportunities to avail themselves of family planning services, but in many urban areas fertility has remained high.

All countries of the world appear to share in the increase in the number of elderly, but the proportion of elderly has only increased dramatically in industrialized societies. In many Third World areas, the proportion of young is as high as 45–50 per cent, and within the next ten years this is unlikely to change significantly.

The increased number of people in many parts of the world, and the changing composition of populations everywhere, pose challenges of many sorts for development activities. For several decades at least, population growth, and the distributional and compositional effects of growth, will remain a significant component of the development scene.

References

Alam, I. & R. Leete 1993. Variations in fertility in India and Indonesia. In *The revolution in Asian fertility*, R. Leete & I. Alam (eds), 148–72. Oxford: Oxford University Press.

Blacker, J. G. C. 1993. Trends in demographic change. *Royal Society Tropical Medicine and Hygiene, Transactions* 87, 3–8.

Bongaarts, J. & R. G. Potter 1983. *Fertility, biology and behaviour: an analysis of the proximate determinants.* New York: Academic Press.

Caldwell, J. 1986. Routes to low mortality in poor countries. *Population and Development Review* 12, 171–220.

Coale, A. J. & P. Demeny 1983 (1966). *Regional model life tables and stable populations,* 2nd edn. Princeton, NJ: Princeton University Press.

Easterlin, R. 1975. An economic framework for fertility analysis. *Studies in Family Planning* 6, 54–63.

Garnett, G. P. & R. M. Anderson 1993. No reason for complacency about the potential demographic impact of AIDS in Africa. *Royal Society of Tropical Medicine and Hygiene, Transactions* 87, 19–22.

Garfield, R. & G. Williams 1989. *Health and revolution: the Nicaraguan experience.* Oxford: Oxfam.

Gwatkin, D. 1979. Political will and family planning. *Population and Development Review* 5, 29–60.

John Hopkins University 1980. *Complications of abortion in developing countries*. Series F (7), Population Information Program, John Hopkins University.

IPPF 1993. *Meeting challenges: promoting choices*. A report on the 40th anniversary IPPF Family Planning Congress, New Delhi, India. Carnforth, England: Parthenon

Langford, C. & P. Storey 1993. Sex differentials in mortality early in the twentieth century: Sri Lanka and India. *Population and Development Review* 19, 263–82.

Lapham, R. J. & G. B. Simmons (eds) 1987. *Organizing for effective family planning programs*. Washington DC: National Academy Press.

Larsen, U. & J. Menken 1991. Individual level sterility: a new method of estimation with application to sub-Saharan Africa. *Demography* 28, 229–60.

NCPD/IRD 1989. *Kenya demographic and health survey 1989*. Nairobi: National Council for Population and Development/ Columbia, Maryland: Institute of Resource Development-Macro Systems.

The Observer 1989. The "silent genocide" of millions of children (report by Annabel Ferriman, Health Correspondent, 1 October).

Peng, X. 1991. *Demographic transition in China*. Oxford: Oxford University Press.

Phillips, J. F. & J. A. Ross 1992. *Family planning programmes and fertility*. Oxford: Oxford University Press.

Population and Development Review 1993. The Dakar Declaration on population. *Population and Development Review* 19, 209–15.

Robey, B., S. O. Rutstein, L. Morris 1993. The fertility decline in developing countries. *Scientific American* 269(6), 30–37.

Thomas, I., J. Cameron, T. Lusty, C. Walker 1987. *UK Overseas Development Administration population programme for Pakistan, 1988–1993*. A report of a pre-appraisal mission visit to Pakistan, Overseas Development Group, University of East Anglia.

UN 1984. *Kilimanjaro programme of action for African population and self-reliant development*. Arusha: Government of Tanzania / UN Economic Commission for Africa.

— 1988. *Mortality of children under age 5: world estimates and projections 1950–2025*. New York: United Nations.

— 1989a. *World population at the turn of the century*. New York: United Nations.

— 1989b. *Levels and trends of contraceptive use as assessed in 1988*. New York: United Nations.

— 1990 (annually thereafter). *Human development report*. Oxford: Oxford University Press.

— 1991. *World population prospects 1990*. New York: United Nations.

UNICEF (annually). *The state of the world's children*. Paris: UNICEF.

Wolfson, M. 1978. *Changing approaches to population problems*. Paris: OECD / World Bank.

World Bank 1990. *World development report 1990: poverty*. Washington: IBRD / Oxford: Oxford University Press.

— 1993. *World development report 1993: investing in health*. Washington: IBRD / Oxford: Oxford University Press.

INDEX OF PERSONS NAMED

SUBJECT INDEX

211

Printed in the United States
by Baker & Taylor Publisher Services

Printed in the United States
by Baker & Taylor Publisher Services